CONSUMER GUIDE®

THE FOOD PRESERVER

By the Editors of Consumer Guide®

Publications International, Ltd.
Skokie, Illinois

Manufactured in the United States of America
1 2 3 4 5 6 7 8 9 10

Library of Congress Cataloging in Publication Data

Main entry under title:

Consumer guide the food preserver.

1. Canning and preserving. 2. Food, Frozen.
I. Consumer guide. II. Title: The food preserver.
TX601.C64 641.4 75-35690
ISBN 0-671-22227-9
ISBN 0-671-22228-7 pbk.

Trade distribution by Simon and Schuster
A Gulf + Western Company
Rockefeller Center, 630 Fifth Avenue
New York, New York 10020

Acknowledgments:
The Editors of Consumer Guide Magazine wish to award a blue ribbon thank-you to all the champion food preservers who sent in their winning recipes. We feel we have made many friends throughout the United States and we are happy to share their enthusiasm for making jellies, jams, pickles and relishes.

Our special thanks goes to the many Cooperative Extension Service specialists and State and county fair supervisors who put us in touch with the blue ribbon winners.

Cover design: Frank E. Peiler
Illustrations by Sandi Zimnicki

Contents

Blue Ribbon Food Preserver

Good news! Our heritage of tasty, thriftily preserved foods, is alive and thriving. As proof, *The Food Preserver* updates the basic preserving methods and proudly shares blue ribbon recipes from across the United States. These champion blue ribbon recipes for jam, jellies, pickles and relishes won prizes at State and county fairs for beginners and experienced cooks — men, women and youths. They represent a criss-cross of regional specialties and classic recipes: from hot Pickle Peppers made in Ringgold, Georgia, to Crab Apple Jelly from Blooming Prairie, Minnesota; from Dutch Salad made in Presque Isle, Maine, to Grandma's Mock Mincemeat from Beavercreek, Oregon.

The Food Preserver shows you how to be a blue ribbon winning food preserver, too. Whether you have a tiny kitchen in a city apartment, or plan to "put up" foods for a large suburban or farm family, you will find the basic steps clearly illustrated for canning, pickling, preserving, freezing, drying, smoking, and cold-storage.

It is not just nostalgia that is sending more and more people back to making their own pickles, relishes, jams and jellies. They taste better made at home. Also, the desire to pare down food prices and escape the impersonal mechanism of commercially processed foods has turned people back to "old-time" methods like smoking, drying, cold-storing, as well as canning and preserving. *The Food Preserver* modernizes traditional preserving methods with step-by-step instructions that beginners can readily follow and experts can use to make their preserving more efficient. The freezer, the great 20th Century contribution to food preservation, has not been overlooked, either. Keeping fruits, vegetables, meats, poultry and fish fresh-tasting and attractive is easy when you follow *The Food Preserver's* directions.

Each chapter lists the basic equipment and basic ingredients you will need. Then, the illustrated instructions take you through a basic recipe, step by step. The basic steps are followed by a treasury of recipes. The blue ribbon recipes appear in the chapters on pickling and making jellies and jams. Each chapter is prefaced with a bit of history about food preservation in the American colonies and into the 20th Century. Many of the recipes, as well as methods, have family trees with roots in Colonial times and beyond. After all, preserving methods, other than canning, have been developing since ancient times.

Winning Blue Ribbons

Blue ribbons, or purple champion ribbons, hanging from jars of peaches, jams or vegetables, used to be the exclusive pride of farm women; but, now everyone is entering these traditional competitions. A blue ribbon from the oldest State fair in the United States was won by a young man from Syracuse University for his third attempt at jam-making.

So, beginners can win blue ribbons, too. It is simply a matter of following directions and knowing the standards used by the judges. In each section of this book, we introduce you to the basic steps of food preservation. You will find out what age and type of food to use for the best results, and what the finished product should look like. Judges check for exactly the same things: size, shape and maturity of food, color, particular standards for particular methods, and flavor.

Entering a Fair

If you want to enter a canned goods competition, first get the Premium Book or entry information from the fair or competition office and read it carefully. We watched a perfect quart jar of peaches passed over because the entry required pint jars. Note the type of pack as well as the jar size. Also note if there is any date specified for the food; early summer fairs may judge the previous summer's tomatoes, for example.

Thanks to the Blue Ribbon Winners

Our editors sent letters to recent blue ribbon fair winners across the nation. The recipes sent to us in response were then tested and written in a standard form for publication. We extend our thanks to the blue ribbon winners who wanted to share their recipes. The American tradition of superior homemade jams, jellies, pickles and relishes will not succumb to commercial products as long as there are organizations like 4-H and individuals like the blue ribbon winners in this book.

For future editions of *The Food Preserver,* we extend an invitation to blue ribbon winners to send their winning recipes to us for publication. Please mention in which fair you entered your recipe and what prize it won and the yield the recipe makes. Write to:

Blue Ribbon Recipes
CONSUMER GUIDE Magazine
3323 West Main Street
Skokie, Ill. 60076

Which Preserving Method?

Which is best? Canning, freezing, or pickling? The answer is a firm and definite, "It depends." It depends on where you live, how you live, what kind of storage space you have, how big your family is, what type of food you want to preserve, if you have a garden, how much time you have to prepare food for preserving, and how much time you have to heat and serve preserved food. Each method has its advantages and disadvantages. You must decide which advantages outweigh the disadvantages for your particular needs.

Canning is a relatively inexpensive method of preserving most fruits and vegetables the second year that you do it. Why the second year? Because you have already made the initial outlay for equipment (water bath or pressure canner, jars). Once that beginning expense is past, canning can be a cheap way to put food by — if you have a garden or an inexpensive source of quality produce in quantity; if you have plenty of shelf space where it is cool, dark and dry; and if you have shelf space to store the bulky equipment needed for canning. You probably will not save much money the first time you can, but plan and plant wisely and you can harvest savings in the following seasons.

A recent study by Cornell University nutritionists proved that savings can be considerable if you grow your own produce and if you already own jars. Their figures price a home-canned quart jar of tomatoes (jar already owned) at 4.3 cents. This same quart cost 50.9 cents if the jars and tomatoes were purchased. A commercially canned quart of tomatoes cost 62 cents to 78 cents at the time of this study. All home-canned food should be brought to a boil, covered and boiled 15 minutes before tasting or serving.

Making jam, jelly, pickling and preserving are all forms of canning. Even a tiny apartment kitchen can handle small batches of jam, jelly, pickles or relishes. An excellent way to preserve a variety of fruits and vegetables, the results are usually far superior to their store-bought counterparts. Many of the recipes for jams, pickles and preserves are not available in stores; they make excellent gifts, too. Properly made jams, jellies, pickled fruits and vegetables and relishes do not need to be boiled before tasting.

Freezing is the most convenient method of food preservation. Frozen foods usually taste better and retain more nutrients than food preserved by any other method. Freezing takes little in the way of equipment: a blancher, some containers, packaging materials. But the freezer itself is a big investment, and shelf space in a freezer costs considerably more than shelf space in a closet for canned goods. The freezer itself takes up space. You may decide that convenience, and saving time, do have some dollar value and that a sub-zero chest full of main dishes, fruits, vegetables, desserts and miscellaneous foods is a better deal than rows and rows of jars of tomatoes, peaches or green beans.

The Cornell study showed that it cost about 19 cents per pound of food to freezer-store it for a year. This cost includes the initial investment in a freezer, electricity, packaging materials, repairs and depreciation. The study proved that it was much more efficient and economical to buy a large freezer, to use it regularly and turn over the inventory often than to use a small freezer for long term storage with little in and out activity. The figures were 3.3 cents per pound of food kept in large, often-used freezers, as compared to 9.6 cents per pound of food for small, infrequently used freezers.

Drying is a simple way to preserve fruits, vegetables and herbs. If you live in a hot, dry part of the country, all you have to do is prepare the food, put it out in the sun and keep away the bugs. Drying is not quite as easy in areas that do not have climates like California or Arizona. You will have to rely on an oven or box dryer. You will have to pay for the energy used and be on hand to turn and stir the foods as they dry.

Your life-style determines how much food you need, how much you can grow and how much you want to put away. If you have a garden and a big family you may want to try all methods; canning tomatoes and green beans; freezing meats, poultry, fish, combination main dishes, pies, bread; pickling cucumbers and peaches; making grape and apple jelly. The foods you like best are the ones you will want to preserve.

The keys to success with each method are the same:

- Select high quality food.
- Handle it quickly and carefully.
- Follow the directions, methods and recipes in this book to the letter.

Lids

Last year's lid shortage sent canners scurrying for substitutes or to other preservation methods. As this book goes to press, lid manufacturers assure us there will be plenty of the standard, flat metal lids available. Several new manufacturers have entered the field, and the two long-time manufacturers have increased production, one by 50 percent.

Some of the new lids available for the 1976 canning season will use new sealing methods. These new lids have not been available for testing since they are still in production. We recommend that you stick with the standard two-piece self-sealing lids, rimmed with sealing compound and held in place by a metal screw band.

Safe Food Handling

Did you know that your case of "24-hour bug" might actually have been caused by something you ate? It is true. Food-borne illness is surprisingly common, and it is caused by poor handling practices in the kitchen.

Food cleanliness and safety are important for every kitchen job. Cleanliness and safety are even more important in food preservation, where microorganisms have the time, and the right conditions, to grow and do their dirty work.

For clean and safe-to-eat food, always follow the recipe directions exactly, use containers and equipment called for, and keep in mind the Do's and Don'ts that follow:

Do

1. Work in a clean kitchen, with clean floors, countertops, cabinets and range, and clean equipment.
2. Keep utensils that handle raw meat and poultry scrupulously clean. This means scrubbing, washing and rinsing knives and cutting board between each type of cutting or chopping task.
3. Work with clean food. This means washing, scrubbing and rinsing fruits and vegetables in several waters, lifting food out of the water to drain and not letting water drain off over food.
4. Work with clean dish cloths and towels, clean hands and clean clothes.
5. Wash your hands each time you touch something other than food, such as your hair, face, the phone, a child or a pet.
6. Get out all equipment, wash and get it and all ingredients ready before you start to follow recipe directions so there will be no delays, no chance for food to spoil.
7. Remember to protect your hands when working with hot foods and jars. Use hot pads, tongs, jar lifters.
8. Be extra cautious with large pots or kettles of boiling water or food. Don't move them, but keep them on the range and work there.
9. Avoid any sudden changes in temperatures when working with hot jars of hot food. Putting a hot jar on a counter or in a cold draft could break it.

Don't

1. Do not use your hands when a kitchen tool will do the job. Keep fingers out of food if at all possible.
2. Do not try cuts or substitutions or time-saving gimmicks. There is only one way to prepare food for preserving — that is the right way, and all the techniques and recipes that follow are the right way. Do not change them.
3. Do not cook or prepare food for preserving if you are sick. You do not belong in the kitchen.
4. Do not prepare food if you have sores on your hands, unless you wear rubber gloves.
5. Do not use home-canned foods without checking carefully for signs of spoilage — bulging or unsealed lids, spurting liquid, mold or off-odor. Discard the contents of any jar showing these signs. Be sure you dispose of the spoiled food where no humans or animals can accidentally eat it.

Weights, Measures and Volumes

1 teaspoon = 5 milliliters
3 teaspoons = 1 tablespoon = 15 milliliters
1 tablespoon = 15 milliliters
4 tablespoons = ¼ cup = 60 milliliters
5 ¹/₃ tablespoons = ¹/₃ cup = 79.95 milliliters
8 tablespoons = ½ cup = 0.12 liters
10²/₃ tablespoons = ²/₃ cup = 0.16 liters
16 tablespoons = 1 cup = 0.24 liters
1 cup = 8 fluid ounces or ½ pint = 0.24 liters
1½ cups = 12 fluid ounces or ¾ pint = 0.36 liters
2 cups = 16 fluid ounces or 1 pint = 0.48 liters
1 pint = 0.48 liters
4 cups = 32 fluid ounces or 2 pints or 1 quart = 0.96 liters
2 pints = 1 quart = 0.96 liters
1 quart = 0.96 liters
2 quarts = ½ gallon = 1.92 liters
4 quarts = 1 gallon = 3.84 liters
1 gallon = 3.84 liters
1000 milliliters = 1 liter

Canning

Thanks to Napoleon, the canning process was established. Napoleon's troops needed to be fed and they were far away from supplies. So, in 1795, the Directory offered 12,000 francs to the first person to invent a method of preserving food that could travel with the troops. Nicholas Appert, a confectioner, began experimenting. He packed food in bottles, sometimes cooking the food slightly first. Then he corked the glass containers and put them in a boiler and cooked them for several hours. When the boiling was over, he sealed the corks with wax. Napoleon himself gave Appert his 12,000 francs. Much later in the 19th century, another Frenchman, Louis Pasteur, explained why Appert's process worked.

Basically, the same boiling water bath process is used in home canning today. But Appert's method was used by the military and in commercial industry before it was widely used by housewives.

The British and Dutch were busy solving the food preservation problem, too. If the British had been a few decades faster, Major General Cornwallis probably would not have had to complain about his troops' provision problems. And, with better fed troops, who knows what might have happened?

A patent was taken out by Englishman Peter Durand in 1810 for a process similar to Appert's. Durand's process used tin-plated, sheet steel cannisters — hence the term canning. The British, by a twist of linguistic fate, call canned foods, tinned food. In America, the canning industry was well established by the time the North and the South needed to feed their boys in blue and grey. Underwood, Borden, Ball, Mason, Heinz, Van Camp are all names of American pioneers in the canning process — their commercial success story is obvious from any grocery store.

Closing food in a container and heating it for long-term storage probably came before Appert and Durand, though the process was not used with any regularity. Our own first author of a cookbook, Amelia Simmons, offers a method for preserving plums in bottles by boiling the bottles. The main drawback to home canning was the lack of canning equipment. Glass was scarce, so was pottery, and tin was something that was recycled from its original use — like holding snuff.

Commercially canned food was a luxury item throughout the 19th century. In rural 19th-century America, it was the canning method, rather than commercially canned products, that eventually appeared in the kitchen.

Hot and Cold Pack

There are two ways to pack food into jars, called the hot pack method and the cold pack method. The cold pack or raw pack is just what the name implies. You prepare fruits or vegetables by peeling, pitting, cutting; pack them into jars uncooked; and then cover them with boiling and processed liquid. For fruits the liquid may be syrup or juice; for vegetables, water is usually used. Since uncooked foods shrink slightly because they are cooked by the processing, you must pack them firmly into jars.

Hot pack foods are briefly pre-cooked before you put them into jars. Cooked foods can be packed more loosely into jars. Processing is often shorter for hot packed foods, since they are already heated through when they go into the canner.

Processing

When you heat food sealed in jars to high temperatures for specific times, you are processing. Processing is absolutely necessary for all fruits and vegetables, as well as pickles, jams, jellies and condiments. Processing kills the microorganisms that could cause spoilage.

Fruits are processed in a boiling water bath, to temperatures of 212°F. Vegetables, meats, poultry, soups and so forth are processed in a pressure canner to 240°F. Descriptions of boiling water bath and pressure canners follow, along with step-by-step procedures for each type of processing.

Do not try to short cut either of these methods, or substitute gimmicks such as preserving powders, aspirin, dishwasher-canning, oven-canning or microwave oven canning. You tempt trouble — spoiled food, sickness, even death.

Do not use old-fashioned recipes that call for open kettle processing. This method has you cook food in an open kettle, then pack it into jars and seal it without any processing. Because of the lack of processing, the chances of spoilage are great.

Botulism

Botulism is a disease caused by the toxin produced in a sealed, low-acid environment, by *Clostridium botulinum.* This bacteria is found just where you find most vegetables — in the dirt. A sealed canning jar is the perfect place for this bacteria to thrive. You can kill it if you heat the food to 240°F. You can heat food to 240°F by processing it in a pressure canner, where steam (hotter than water) provides the heat. All vegetables must be processed in a pressure canner.

If you follow procedures and processing times exactly and if your pressure canner is in good operating condition, the food should be safe to eat. As a further precaution, we recommend that you bring home-canned food to a boil, cover and boil 15 minutes before tasting or serving.

Getting Organized

Once you have learned the procedures and precautions of canning it is then just a matter of getting it all together. If you are about to can for the first time, please read the rest of this section, including the step-by-step directions, and read the recipe through before you start. If you are an experienced canner, it still pays to refresh your memory at the start of each canning season.

Check recipes and figure out the number of jars you will need. Get out all the other equipment, wash and rinse it. Be sure everything is in working order. Assemble and prepare all the ingredients.

Time

Canning need not take hours and hours out of your day. You can prepare and process food as it ripens in your garden, perhaps putting away a canner-full each day. You should not prepare more food at one time than will fill one group of jars in the canner, anyway. So, organize canning to fit your schedule. Perhaps one canner load in the morning, another in the evening after it has cooled off; perhaps you can devote a whole day to several canning sessions, if that is more convenient. Work fast with small amounts at each canning session.

Do not try to manage anything else while you are canning. It can be difficult but try to avoid delays. This means avoiding other kitchen chores, children and the phone. Keep small children out of the kitchen. Older children can help — preparing food, washing jars, sealing containers — but the little ones may be in danger, and they will cramp your style.

Canning takes ample work space, so plan ahead. You will need:

- Sink room for washing and preparing food.
- Counter space for sorting, chopping or cutting.
- Range space for cooking, processing and heating water.
- More counter space for cooling jars. A sturdy table, set up out of traffic and drafts, makes a good cooling area for jars.
- Shelf space in a clean, cool, dark, dry storage area where the food will not freeze.

If you have canned before, take an inventory to see how much you should put up this year. Gardeners are wise to take inventory before they plan their gardens so they will know how much of what to plant to put up this season. A look at what is left from last year will help determine the size of your crops.

Be sure to move last year's jars to the front of the shelf so they will be used first.

Basic Equipment

Boiling Water Bath Canner

You will need a water bath canner for processing all fruits, as well as pickles and relishes, jams, preserves and conserves. Several well-known pot and pan manufacturers make water bath canners, usually of aluminum. The canner is a big pot (20- to 21-quart capacity) with a lid and a special rack that fits down inside the pot to hold the jars. Often this rack has handles and special ridges so that you can lift the rack out of the water and hold it in place while putting in the jars. These pots are very big so plan your storage space accordingly. You can use the pot for many other big cooking jobs — such as spaghetti and corn on the cob, for instance.

On old ranges, sometimes a big pot can completely cover a burner, shutting off the air to pilot light or blocking the heat so that the enamel surface of the range becomes too hot. Put the canner on the burner of your range to check if it "seals off" the burner. If it does get several blocks of sheet asbestos, about 1/2-inch thick, to raise the canner. Lumberyards or hardware stores should have asbestos.

If you already own a very large pot you can probably turn it into a water bath canner by finding a rack that will fit down in the pot, covering the bottom. Jars must be on a rack — if they stand directly on bottom they may break. The pot must be deep enough to allow a minimum of two inches of space above the jars as they sit on the rack. The jars must be covered by at least one inch of water (two is better) with another inch or two for boiling room. Half-pint and pint jars obviously can go into a shallower pot than quarts. To help you estimate, pint jars need a pot a minimum of 8 to 10 inches deep, quarts a minimum of 9 to 12 inches deep.

You may also use a pressure canner as a boiling water canner. Just place the rack in the bottom, pour in the water and put the jars in place. But, do not lock cover in place and be sure to leave the petcock or vent open so steam can escape.

Pressure Canner

You will need a pressure canner to process any vegetable, meat, poultry or any combination of these foods. A pressure canner can also be used for quick cooking of many foods, as

Boiling Water Bath Canners. Use for fruit, jam, pickles. Must be deep enough for 1 to 2 inches of water to cover jars, plus 1 inch for boiling.

Pressure Canner. Use for vegetables, tomatoes. Some have dial gauges, others have weight gauges. Always read the manufacturer's directions.

the instruction book that comes with a canner explains.

A pressure canner is a large, heavy metal utensil that heats water under pressure to create steam. The steam is hotter than boiling water and can cook food to the 240°F needed to kill dangerous bacteria.

Pressure canners come in several sizes, ranging from 8 to 22 quarts in capacity and holding 4 to 7 quart jars, or many more pint jars of food. You may already have a pressure cooker, or saucepan pressure cooker. You can use this for processing if it will hold pint jars, if it has an accurate gauge and if it will maintain 10 pounds of pressure.

Altitude can affect pressure, so if you live more than 2000 feet above sea level you will need to increase the pressure, adding one pound for each 2000 feet above sea level. If the pressure canner has a weighted gauge, use 15 pounds pressure if you live above 2000 feet.

Pressure canners (and cookers) vary slightly in construction — one brand has a dial gauge, another a weight gauge. Always read and follow the instructions that come with your canner to the letter. Before you start to prepare food to be processed in a pressure canner, take out your canner and book and become acquainted with all parts, controls and instructions. (If you have lost the instruction book, write to the manufacturer for another, giving the model number of your canner.) Study the base, handles, and rack that fits inside, then look over the cover carefully, noting the dial or pressure control and vents. Check the gasket and locking mechanism.

Dial gauges must be checked each canning season. You can send the dial to the manufacturer for testing, but it may be easier to take it to your local extension office, where home economists have set up testing facilities. Just call beforehand to be sure they are ready to test. Weighted gauges do not need to be tested, but they must be kept clean. Handle the dial gauges with care — never rest the cover on the gauge and never turn the cover upside down over a full pan with the gauge attached because moisture could enter the gauge and rust it.

Just as with large water bath canners, big pressure canners may cover the range burners completely. Set the pressure canner on the burner to be sure there is enough air space to keep gas burners on or to prevent the enamel of the range surface around the electric unit from growing too hot. If a pressure canner seems too snug against a burner, lift it up a fraction of an inch on asbestos blocks (from a lumberyard or hardware store), so that heat and air can circulate.

Remember to become acquainted with your pressure canner before starting to use it. Familiarity breeds confidence and dispels fear!

Jars and Lids

Standard 1-pint, 1½-pint, or 1-quart jars with two-piece self-sealing lids are the only proper containers for canning fruits, vegetables and all other foods. The recent lid shortage brought out some new and some old types of lids and closures, but we recommend that you stick with the standard flat metal lid, rimmed with sealing compound, and the accompanying metal screw bands to hold the lids in place for sealing and processing.

Do not try to use jars other than those especially made for canning. Peanut butter, mayonnaise, instant coffee or other food jars are not tempered to withstand the heat of processing, and their top rims may not be right for the lids. Do not risk losing food, breaking jars and getting cut by using substitutes for canning jars.

Always check standard glass canning jars for each use, to be sure they are free from nicks and cracks. Run your finger around the top of each jar — it should be flawless. If it is not, do not use it for canning.

Use only standard glass canning jars. Select 1-pint, 1 1/2-pint or 1-quart jars with 2-piece self-sealing lids.

Jars and screw bands can be used year after year. Lids cannot be reused. In fact we recommend you open sealed jars with a pointed opener, piercing the lid so there is no question about its age.

Additional Basic Equipment

1. **A wide-mouth funnel and ladle** with a pouring lip make jar filling neat and easy. Many funnels sit down ½-inch inside the jar so you can use the bottom for a head space guide.
2. **A large preserving kettle, saucepan or pot** will be needed for most hot packed foods. Do not use iron, copper or tin pans — they may discolor food or cause bad flavors.

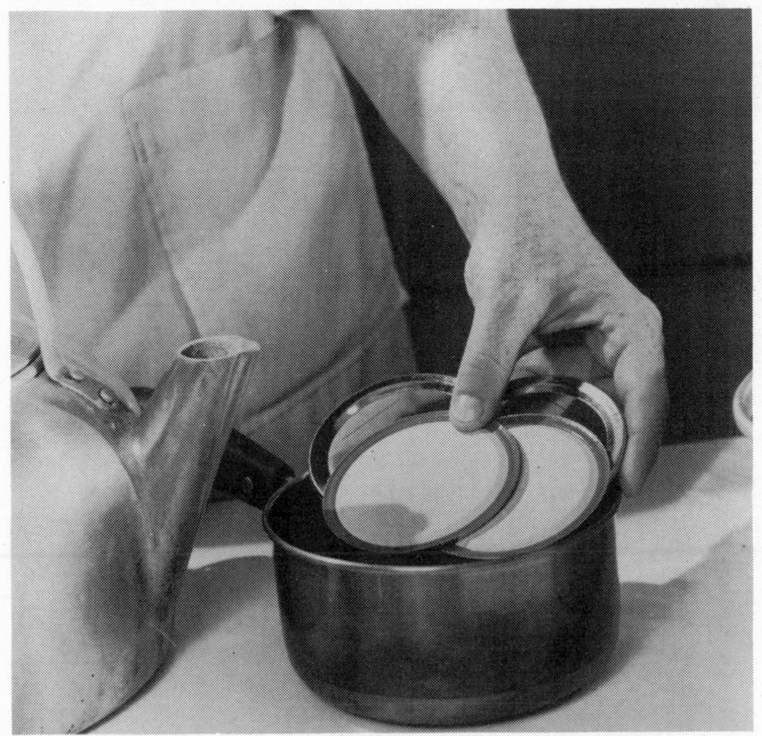

Prepare lids as the manufacturer directs. Some require boiling water to be poured over them.

3. **Spoons** will be needed for stirring, spooning and packing. Use wooden spoons for stirring, slotted spoons for lifting, and smaller spoons for filling. Accurate measuring spoons are essential.
4. **Knives** are a necessity. A sharp chopping knife or chef's knife and a good paring knife will handle most tasks.
5. **Long-handled tongs and/or special jar lifters** are not expensive and they are very helpful for taking jars in and out of boiling water. Do not try to manage without these helpers.
6. **Measuring equipment** should include both dry (metal) and liquid (glass) measuring cups. It is handy to have a full set (¼, ⅓, ½ and 1 cup) dry measures along with a 2-cup measurer. One-cup, 1-pint and 1-quart liquid

Additional Basic Equipment. Always read a recipe through and check your equipment and ingredients before you start to can.

measurers will simplify canning. Household scales, that can weigh from ¼ pound (4 ounces) up to 10 or 25 pounds, help you measure produce.

7. **A timer** saves you from clock-watching cooking and processing times.
8. **A strainer or colander** helps hold fruit after it has been washed or rinsed, and may also be necessary for draining cooked fruit.
9. **A teakettle or large saucepan** will boil the extra water you may need to cover jars in a boiling water bath, to fill a pressure canner or to cover vegetables in jars.
10. **Hot pads, oven mitts, wire cooling racks or folded dish towels** protect your hands and countertops from the hot jars. Some canners keep a pair of old cotton gloves clean and at hand to wear while filling and sealing hot jars.
11. **A nonmetal spatula** or tool with a slim handle, such as a rubber spatula, wooden spoon, or even a plastic knife, is needed to run down the sides of filled jars to release air bubbles. Metal knives or spatulas could nick the bottoms of the jars.

Basic Ingredients

Each recipe tells you what to look for when picking or buying produce. Sort by size and shape and maturity, then handle similar items together. They will look better in the jars and cook evenly.

Freshness is extremely important. Once foods are picked they begin to lose flavor, and they can start to spoil quite quickly. Rush foods from the garden to the kitchen and into the jars. Experts say two hours from harvest to jar is the best time span to capture flavor. If you cannot prepare foods right away, be sure to refrigerate them immediately but do not let them languish in the refrigerator for too long.

Canning for Special Diets

Unlike pickling and preserving, sugar and salt are not crucial to the canning process. They are only added for flavor. If you have a special dieter in your family, just omit the sugar and/or salt. Or you may use part honey or light corn syrup for part of the sugar when preparing syrup for fruit.

Processing Times

The times in the recipes that follow are given for both pints and quarts. If you use ½-pint jars, follow times for pints; if using 12-ounce or 1½-pint jars, process as for quarts. Never skimp on processing times.

Nobody's Perfect

There are so many factors involved in canning that sometimes things can go wrong. Here are some common problems, causes, and solutions.

1. **Spoilage** is caused by improper handling, underprocessing or faulty seals. Follow all directions carefully, use the proper equipment and the right kind of food, as directed in the recipes. The signs of spoilage are:
   Bulging lids
   Broken seals
   Leaks
   Change in color
   Foaming
   Unusual softness or slipperiness
   Spurting liquid (when the jars are opened)
   Mold
   Bad Smells

   If you notice any of these signs, do not use the food. Discard it where animals and humans cannot find it. You can salvage the jars by washing them well, rinsing and then boiling them for 15 minutes.
2. **Jars can lose liquid** during processing because they were filled too full and liquid boiled out, because air bubbles were not released, because pressure fluctuated during processing, or because you tried to hurry the pressure reduction and liquid was forced out of jars by the sudden change in pressure. Do not try to add more liquid to jars — the food is safe as is. If you add more liquid you must put food in clean, hot jars, seal with new lids and reprocess.
3. **Underprocessing** can be caused by skimping on processing time, not having an accurate gauge on pressure canner, not exhausting canner as manufacturer directs, not following headspace guides (especially for foods that ex-

pand during processing — corn, peas, lima beans).

4. **Jars that do not seal** could be the result of flaws in jar or lids, lack of heat, or food on the rim of the jar. The seal could have been broken by tightening the screw band after removing the jar from the canner or by turning the jar over as it is removed from the canner. Food in jars that do not seal can be repacked immediately in clean jars with new lids and reprocessed. Or, if just a jar or two fails to seal, refrigerate the contents and use right away.
5. **Fruit that floats** in cold packed jars can be the result of overcooking, too little fruit for the amount of liquid, too heavy syrup or because fruit has spoiled. Check carefully for other signs of spoilage.
6. **Food in jars discolors** because the jars may not have been filled full enough (air at top of jar caused food to darken), or because processing time was too short. Iron or copper in the water, or storage in light can also cause discoloration. Some varieties of corn, and overmature or overprocessed corn can also discolor. Color changes do not always mean that food has spoiled, although spoiled food may be discolored. Check for other signs of spoilage.
7. **The underside** of the jar's lid may discolor from chemicals in the food or water. This is common and nothing dangerous.
8. **Jars that break** in a canner probably had hairline cracks before they were filled. The jars may have bumped each other during processing, or were placed directly on the bottom of canner without a rack to protect them.
9. **A scum or milky powder** on the outside of jars, noticeable after processing or cooling, is just from minerals in the water. Wipe it off with a cloth and do not worry about it. Next time, add a tablespoon or two of vinegar or a teaspoon of cream of tartar to the water in the canner. This also helps prevent staining inside of the canner.

Canning Fruit

Memories of a pantry lined with canned peaches, apricots and cherries linger on. Pantries may have been eliminated from modern houses, but canning fruit continues to be one of the most popular preserving methods. Besides recipes for favorite summer fruits — from apples to rhubarb — you will find directions for canning applesauce, fruit cocktail and fruit juices in this chapter. Be sure to review the Basic Steps for Canning Fruit.

Hot and Cold Packs
Most fruits may be hot or cold packed. Many of the recipes that follow give both methods. All fruits must be processed in a boiling water bath for the precise times given in the recipes.

Basic Equipment

The basic equipment for canning fruit includes all the Basic Equipment for Canning, except a pressure canner. Fruit is processed in a boiling water bath canner. Check Basic Equipment for Canning for a detailed description of the following necessary equipment.

Boiling water bath canner
Standard jars (1 pint, 1½ pint, 1 quart) with 2-piece self-sealing lids
Preserving kettle (for the hot pack method)
Teakettle
Strainer
Knives — for paring and chopping
Measuring cups
Measuring spoons
Spoons — wooden and slotted
Wide-mouth funnel and ladle
Jar lifter or tongs
Timer
Hot pads
Nonmetal spatula, plastic knife or other slim handled tool
Wire cooling racks or folded dish towels

Basic Ingredients

Use fruits that are as ripe and perfect as possible — no spots, decay or softness from storage. Apples should be crisp and flavorful, not mushy and bland. Apricots, nectarines, peaches, pears, pineapples and plums should be ripe, but firm. Check berries to make sure they are free from mold. Cherries should be tree ripened and unblemished. Cranberries should be firm, without bruises. Choose tender rhubarb stalks that have good color.

Basic Equipment for Canning Fruit. Organize your work area and all the equipment before you start to can.

Basic Ingredients for Canning Fruit. Besides sweetener for syrup and water, you will need perfect fruit, sorted by size and ripeness.

Syrup and water are the only other ingredients used in canning fruit. White granulated sugar, light corn syrup or mild-flavored honey can be used to make the syrup.

Treat To Prevent Darkening

Apples, peaches and pears need to be peeled, then treated to prevent darkening. There are several possible treatments; pick the one that is most convenient for you. As you peel fruits you can drop them in a solution of salt water and vinegar (2 tablespoons salt and 2 tablespoons vinegar in 1 gallon cold water), or drop the fruit in water with commercial ascorbic acid mixture added (3 tablespoons of commercial ascorbic acid to 2 quarts of water). Drain when ready to pack the fruit into the jars. Drug stores sell ascorbic acid, also known as vitamin C.

Other ways of preventing darkening are to add 5 (50 milligram) tablets of ascorbic acid to each quart jar, or sprinkle ⅛ to ¼ teaspoon powdered or crystalline ascorbic acid over the fruit in the jar just before sealing. You can also add ascorbic (½ teaspoon powder or crystals) or commercial ascorbic acid mixture (4 teaspoons) to each quart of syrup just before pouring the syrup over fruit in the jar.

Sweetening Fruit For Canning

Most fruits can be packed in syrup. Some juicy fruits keep their shape and are more flavorful if they are packed with granulated sugar. To use sugar, just heat each quart of prepared fruit with ½ cup sugar. Bring it to a simmer, then pack the fruit in the juice that remains after simmering. If you do not have enough syrup to completely fill the jars, use boiling water.

You can replace part of the sugar with light corn syrup or mild-flavored honey. Do not use brown sugar, molasses, dark corn syrup or strong-flavored honey or you will overpower the delicate fruit flavors.

Unsweetened Fruit

There is no need to use sugar if you are watching calories or have a special dieter in the family. Use noncaloric sweeteners, following manufacturer's directions, or pack the fruit in its own juice, in other fruit juice or just in water. To extract juice for packing, crush or chop ripe fruit, heat the fruit to simmer with just enough water to prevent sticking. Strain the cooked fruit through a jelly bag or cheesecloth. Discard the pulp.

Syrup For Canning Fruit

To make syrup, combine water and sugar (or part sugar and corn syrup or honey) in the proportions given below and boil for 5 minutes. Skim off any scum. Keep the syrup hot, but do not let it boil down. Count on 1 to 1½ cups of syrup for each quart jar. Leftover syrup can be refrigerated to use another time.

Type syrup	Sugar cups	+	Water cups	=	Syrup cups
30% (thin)	2		4		5
40% (medium)	3		4		5½
50% (heavy)	4¾		4		6½

Basic Steps

1. Select fruits that are perfect, free from blemishes or decay and completely ripe. Sort by size and ripeness and handle the ones that are alike together.
2. Set out all the ingredients and equipment. Wash and dry all the equipment, countertops, working surfaces and hands. Wash and rinse the jars, keep them hot in water, a low oven or the dishwasher on dry cycle. Prepare the lids as the manufacturer directs.
3. Wash fruits thoroughly in plenty of water, with several changes of water and rinses. Handle the fruits carefully to avoid bruising. Scrub firm fruits with a brush. Always lift fruits out of water rather than letting water drain off over the fruit, redepositing dirt. Do not soak fruits.
4. Put the water bath canner on the range and fill with about 9 quarts of water (or 4 inches deep) for pint jars, 10 quarts (or 4½ inches deep) for quart jars. (These amounts are for a 20- to 21-quart canner.) Start heating water. At the same time put a teakettle or large saucepan of water on the stove to heat.
5. Prepare the type of syrup needed, depending on the recipe. Keep the syrup hot but do not let it boil down.
6. Prepare the fruit as the recipe directs, cutting and peeling only enough for one canner load at a time.

Step 4: Start heating water in the water bath canner. Also start heating extra water in a teakettle or large saucepan in step 12.

Step 5: Prepare the type of syrup called for in the recipe. Keep syrup hot, but do not let it boil down.

Step 6: Peel and halve the fruit and treat as the recipe directs to prevent darkening. Prepare only enough fruit for one canner load at a time.

Step 7: Pack fruit in jars to within 1/2 inch of the top. Cold packed fruit should be firmly packed. Hot packed fruit can be looser.

Step 8: Pour in the hot syrup to within 1/2 inch of the top of the jar. A wide-mouth funnel simplifies packing as well as pouring the syrup.

7. Pack the fruit into jars to within ½ inch of the top. If you follow the cold pack method the fruits should be gently but firmly packed into jars. Fruits for the hot pack method can be packed more loosely since they will not shrink as much during processing.
8. Pour in the hot syrup to within ½ inch of the top of the jar.
9. Run a nonmetal spatula or other slim tool down the side of the jar to release any air bubbles. Pour in additional hot syrup (or juice or boiling water), if needed, to bring level up to within ½ inch of the top of the jar.

Step 9: It is important to release any air bubbles. Run slim, nonmetal tool down the side of each jar. Add more syrup, if necessary.

10. Wipe off tops and threads of jars with damp cloth.
11. Put on lids and screw bands as manufacturer directs.
12. Carefully lower the jars, using long handled tongs or a special jar lifter, into boiling water in the canner. Arrange the jars on the rack so that they do not touch each other or bump against the side of canner. If necessary, add more boiling water, pouring down the side and not directly on the jars, to cover the jars with at least 1 inch of water.
13. Cover and, when water returns to a boil, begin timing for processing. Adjust heat during the processing so that the

Step 10: Wipe off tops and threads of jars with a damp cloth. An oven mitt makes holding the hot jars easier.

Step 11: Put on prepared lids following the manufacturer's directions.

Step 11: Put on the screw bands as the manufacturer directs.

Step 12: Lower jars into boiling water bath using tongs or special jar lifters. Jars should not bump each other or the side of the canner.

Step 12: If necessary, add more boiling water, pouring down the side and not directly on the jars. Cover jars with at least 1 inch of water.

water boils gently but steadily. If the processing time is over 10 minutes, you may need to add additional boiling water to keep the jars covered. If you live at a high altitude you will need to add 1 minute for each 1000 feet above sea level when the processing time is under 20 minutes. If the processing time is over 20 minutes, add 2 minutes for each 1000 feet above sea level. Thus, if you live at 5000 feet add 5 minutes to processing times of 20 minutes or less, 10 minutes to processing times of more than 20 minutes.

14. When the processing time is up, turn off the heat and carefully lift the jars out of the canner. Place the jars several inches apart on a folded towel or rack that is in an out-of-the-way, draft-free place. Do not cover the jars while they are cooling.
15. Let the jars cool for at least 12 hours.
16. When the food in the jars is completely cool, remove the screw bands and check the seals. The lids should be depressed and, when the jar is tipped, there should be no leaks. If the center of lid can be pushed down and comes back up, or if there are any leaks, use the food im-

mediately (store it in the refrigerator) or pour it into another clean, hot jar, seal with a new lid, put the screw band on and reprocess. Do not try to add more liquid to jars that have lost liquid. If you add liquid you have to repack in another clean hot jar and go through the canning process again.

17. Wipe the jars with a clean, damp cloth, then label clearly with the product, batch (if you do more than one a day) and date. Be sure to remove the screw bands — if they are left on they could rust in place. To remove stuck screw bands, wring out a cloth in hot water, then wrap it around the band for a minute or two to help loosen it.
18. Store the jars in a cool, dark, dry place where they will not freeze.
19. Before using, check for signs of spoilage — bulging or unsealed lids, spurting liquid, mold or off-odor — and discard the contents of any jar that shows these signs.

Recipes

The recipes for canned fruits all follow the basic steps and use the Basic Equipment:

Water bath canner
Pint or quart jars and lids
Preserving kettle (for the hot pack method)
Teakettle
Strainer
Knives
Spoons
Jar lifter or tongs
Measuring cups (both metal and glass)
Timer
Hot pads
Nonmetal spatula or plastic knife
Wire cooling racks or folded dish towels

Each recipe will tell which processing method to use — hot or cold pack. Follow the directions carefully, referring back to the Basic Steps, and you will delight your family and impress your friends with rows of delicious canned fruits.

APPLES

Hot pack apples in a 30% syrup, water or fruit juice and enjoy them later in pies, sauces and other desserts. All of the Basic Equipment is used.

2½ to 3 pounds apples = 1 quart, canned
1 bushel (48 pounds) = 16 to 20 quarts, canned

1. Choose crisp, firm and flavorful apples with no bad spots or bruises.
2. Organize and prepare the equipment and work area.
3. Wash, peel and core the apples. Cut into quarters or slices and drop them into salt-vinegar water (2 tablespoons each salt and vinegar to 1 gallon water). Leave them in the salt-vinegar solution until all are peeled.
4. Start the water heating in your water bath canner; start boiling extra water in a teakettle.
5. Prepare a 30% syrup, or use water of fruit juice.
6. Drain, then add enough 30% syrup, water or juice to barely cover the apples. Boil for 5 minutes.
7. Pack the apples into pint or quart jars to within ½ inch of the top. Reserve the liquid, cover and keep it hot.
8. Bring the liquid to a boil and pour it over the packed apples to within ½ inch of the top.
9. Run a nonmetal spatula down the side of each jar to release the air bubbles. Add more of the boiling liquid, if necessary, to fill the jar within ½ inch of top of jar.
10. Wipe the tops and threads of the jars with a damp cloth.
11. Put on the lids and screw bands as the manufacturer directs.
12. Follow Basic Steps for Canning Fruit 12 and 13. Process, in a boiling water bath — 15 minutes for pints, 20 minutes for quarts.
13. Follow Basic Steps for Canning Fruit 14 through 19.

APPLESAUCE

Sweeten and spice this applesauce if you wish, or cook the apples unsweetened. In addition to the Basic Equipment, you will need a sieve or food mill.

2½ to 3½ pounds = 1 quart, canned
1 bushel (48 pounds) = 15 to 18 quarts, canned.

1. Choose tart cooking apples.
2. Organize and prepare the equipment and work area.
3. Wash and core the apples and simmer them with just enough water to prevent sticking until the apples are soft.
4. Start heating water in the water bath canner; start boiling extra water in a teakettle.
5. Press the apples through a sieve or food mill.
6. Sweeten the apples if desired. Use as much or little sweetening as you like. Reheat the apples to boiling.
7. Ladle or pour the sauce into hot pints or quarts to within ½ inch of top.
8. Stir to remove any air bubbles.
9. Wipe the tops and threads of jars with damp cloth.
10. Put on lids and screw bands as manufacturer directs.
11. Follow Basic Steps for Canning Fruit 12 and 13. Process in a boiling water bath — 10 minutes for both pints and quarts.
12. Follow Basic Steps for Canning Fruit 14 through 19.

APRICOTS

You can peel the apricots if you wish, following the directions for canning peaches, but it is not necessary. Use a 30% or 40% syrup. You will need all of the Basic Equipment. Apricots can be processed by either the hot or cold pack method.

2 to 2½ pounds = 1 quart, canned
1 lug or box (22 pounds) = 7 to 11 quarts, canned

1. Choose ripe but firm fruit.
2. Organize and prepare the equipment and work area.
3. Wash the apricots well, drain, halve and pit. Drop them into salt-vinegar water (2 tablespoons each salt and vinegar to 1 gallon water) until all are halved and pitted.
4. Start heating water in the water bath canner; start boiling extra water in a teakettle.
5. Prepare a 30% or 40% syrup.
6. Cold Pack: Drain the apricots and pack them firmly into hot pints or quarts to within ½ inch of top.
   Hot Pack: Drain the apricots and cook a few at a time in boiling syrup until they are hot. Pack them into hot pints or quarts to within ½ inch of top. Pour in boiling syrup to within ½ inch of top.
   Hot Pack: Drain the apricots and cook a few at a time in boiling syrup until they are hot. Pack them into hot pints or quarts to within ½ inch of top.
7. Run a nonmetal spatula down the side of each jar to remove air bubbles. Add additional syrup, if necessary, to fill to within ½ inch of the jar's top.
8. Wipe the tops and threads of the jars with a damp cloth.
9. Put on the lids and screw bands as the manufacturer directs.
10. Follow Basic Steps for Canning Fruit 12 and 13. Process in a boiling water bath — 25 minutes for pints, 30 minutes for quarts.
11. Follow Basic Steps for Canning Fruit 14 through 19.

BERRIES (except Strawberries)

Use a 30% or 40% syrup for the cold pack method. Use sugar for the hot pack method. You will need all the Basic Equipment, except a knife.

1½ to 3 pounds = 1 quart, canned
1 (24-quart) crate = 12 to 18 quarts, canned

1. Choose delicate berries such as red raspberries for the cold pack method. Use blackberries or firm berries for the hot pack method.
2. Organize and prepare the equipment and work area.
3. Wash gently, but thoroughly. Drain.
4. Start heating water in the water bath canner; start boiling extra water in a teakettle.
5. Prepare a 30% or 40% syrup for delicate berries to be cold packed.
6. Cold Pack: Pour about ½ cup of boiling syrup into each hot pint or quart and then fill the jar with berries. Shake the jars to pack the berries. Add more berries and syrup, if needed, to fill to within ½ inch of top of jar.
   Hot Pack: Measure the berries into a kettle and add ¼ to ½ cup sugar for each quart of berries. Let them stand an hour or two to draw out the juices. Then heat until the sugar dissolves and the berries just begin to boil. Pack them into hot pints or quarts to within ½ inch of top. If you run out of syrup, use boiling water to fill jars to within ½ inch of top.
7. Run a nonmetal spatula down the side of each jar to release the air bubbles. Add additional syrup or boiling water, if necessary, to fill to within ½ inch of top of jar.
8. Wipe the tops and threads of the jars with a damp cloth.
9. Put on the lids and screw bands as the manufacturer directs.
10. Follow Basic Steps for Canning Fruit 12 and 13. Process in a boiling water bath — 10 minutes for pints, 15 minutes for quarts.
11. Follow Basic Steps for Canning Fruit 14 through 19.

CHERRIES

Cherries can be processed with the cold or hot pack method. Use 30% or or 40% syrup for sweet cherries, a 40% or 50% for tart cherries. Use sugar for the hot pack method. If you do not have a cherry pitter, use a paper clip or hairpin that you have sterilized by boiling. Besides a cherry pitter, you will need all of the Basic Equipment.

2 to 2½ pounds = 1 quart, canned
1 lug (22 pounds) = 9 to 11 quarts, canned (unpitted)
1 bushel (56 pounds) = 22 to 32 quarts, canned (unpitted)

1. Choose tree-ripened, perfect cherries that have good color.
2. Organize and prepare the equipment and work area.
3. Wash the cherries well and discard any that float. Drain, stem and pit them.
4. Start heating water in the water bath canner; start boiling extra water in a teakettle.
5. Prepare a 30% or 40% syrup for sweet cherries, a 40% or 50% syrup for tart cherries.
6. Cold Pack: Pour about ½ cup of boiling syrup into each hot pint or quart and then fill it with cherries. Shake the jars to pack the cherries. Add more cherries and syrup, if needed, to fill to within ½ inch of top.
   Hot Pack: Measure the pitted cherries and mix ½ to ¾ cup sugar with each quart of cherries. Heat over medium heat until the sugar dissolves and cherries are thoroughly heated. Pack them into hot pints or jars to within ½ inch of the top.
7. Run a nonmetal spatula down the side of each jar to release the air bubbles. Add additional syrup or boiling water, if necessary, to fill to within ½ inch of top of jar.
8. Wipe the tops and threads of the jars with a damp cloth.
9. Put on the lids and screw bands as the manufacturer directs.
10. Follow Basic Steps for Canning Fruit 12 and 13. Process in a boiling water bath. Process the cold pack method: 20 minutes for pints, 25 for quarts. Process the hot pack method: 10 minutes for pints, 15 minutes for quarts.
11. Follow Basic Steps for Canning Fruit 14 through 19.

CRANBERRIES

This recipe is for 4 pints of whole berry sauce or 4 pints of jellied sauce. If you like spiced cranberries, add a stick of cinnamon and a few whole cloves to the berries while they cook. Remove the spices before packing the cranberries into jars. The recipe gives you a choice of whole berry sauce or jellied sauce. You will need a sieve or food mill, in addition to the Basic Equipment, for the jellied sauce. A candy thermometer will help you check the jellied sauce method, but it is not necessary. See the chapter on Testing Jelly for the sheet test and refrigerator test that do not require a thermometer.

2 pounds (2 quarts) = 4 pints, canned whole berries, or 4 pints jellied sauce

1. Choose firm, brightly colored berries.
2. Organize and prepare the equipment and work area.
3. Wash the berries well, pick over and drain. Be sure to pick out the stems.
4. Start heating water in your water bath canner; start boiling extra water in a teakettle.
5. Whole Berry Sauce: Boil 4 cups sugar with 4 cups water for 5 minutes. Add 2 quarts (8 cups) of berries and boil, without stirring, until the skins burst.
   Jellied Sauce: Boil 4 cups berries with 1¾ cups water until the skins burst. Press the berries through a sieve or food mill. Stir in 2 cups sugar and boil to 80°F above boiling (See Testing Jelly for tests that do not require a candy thermometer).
6. Ladle or pour the sauce or jelly into hot pint jars to within ⅛ inch of the top.
7. Wipe the tops and threads of jars with a damp cloth.
8. Put on the lids and screw bands as the manufacturer directs.
9. Follow Basic Steps for Canning Fruit 12 and 13. Process in a boiling water bath for 10 minutes.
10. Follow Basic Steps for Canning Fruit 14 through 19.

CURRANTS (follow the recipe for Berries)

FRUIT COCKTAIL

This delicious fruit cocktail can be served as a salad, appetizer or dessert. You can add maraschino cherries just before serving. You will need a 40% syrup for the fruit and all of the Basic Equipment including a strainer.

Refer the individual fruit recipes in this chapter for the equivalents of raw fruit to canned. The proportion of fruits can vary — you may have a couple pounds of peaches on hand, but only a few pears. Do not prepare more fruit than your canner can handle in one load.

1. Choose firm, but ripe, pineapples, pears, peaches and seedless grapes.
2. Organize and prepare the equipment and work area.
3. Peel pineapples, pears, peaches and then cut them into chunks.
4. Prepare a 40% syrup.
5. Start heating water in the water bath canner; start boiling extra water in a teakettle.
6. Cook each fruit separately in the 40% syrup. Cook until limp, about 3 to 5 minutes.
7. Combine the fruits and ladle them into hot ½-pint or pint jars to within ½ inch of top.
8. Strain the syrup and pour it in the jars to within ½ inch of top.
9. Run a nonmetal spatula down the side of each jar to release air bubbles. Add additional syrup, if necessary, to fill to within ½ inch of the top of the jar.
10. Wipe the tops and threads of the jars with a damp cloth.
11. Put on the lids and screw bands as the manufacturer directs.
12. Follow Basic Steps for Canning Fruit 12 and 13. Process in a boiling water bath — 20 minutes for both ½-pints and pints.
13. Follow Basic Steps for Canning Fruit 14 through 19.

FRUIT JUICES

Use fruit juices for refreshing beverages or, for elegant appetizers, serve "shrubs" made by floating fruit sherbet in small glasses of chilled juice. You will need all the Basic Equipment, plus a cheesecloth or jelly bag strainer.

Use any fruit or combination of fruits that you like. The quantity of juice that is strained out of the simmered fruit is the quantity you will can. Work with small amounts of fruit rather than large amounts.

1. Choose full ripe fruit. Fruits with less than perfect shapes are good for juices. Blemished fruits, with bad parts trimmed out, can also go into the juice.
2. Organize and prepare the equipment and work area.
3. Wash the fruit thoroughly, drain and halve or quarter it. Pit, if necessary.
4. Start heating water in the water bath canner; start boiling extra water in a teakettle.
5. Crush the fruit and heat it just to simmering. Add a little water, if necessary, to prevent sticking.
6. Strain the fruit through a jelly bag or cheesecloth. Discard the pulp.
7. Sweeten the juice to taste, if desired. (About 1 cup sugar to 1 gallon juice should be enough.)
8. Reheat to simmering. Stir until the sugar dissolves.
9. Pour the juice into hot pints or quarts to within ½ inch of the top.
10. Wipe the tops and threads of the jars with a damp cloth.
11. Put on the lids and screw bands as the manufacturer directs.
12. Follow Basic Steps for Canning Fruit 12 and 13. Process in a boiling water bath — 10 minutes for both pints and quarts.
13. Follow Basic Steps for Canning Fruit 14 through 19.

GRAPES

Tight-skinned, seedless green grapes are best for canning. You will need to prepare a 30% or 40% syrup. All the Basic Equipment, except knives, is used.

1 pound = about 1 quart, canned

1. Choose slightly underripe seedless green grapes.
2. Organize and prepare the equipment and work area.
3. Wash the grapes thoroughly, stem and pick over.
4. Start the water heating in the water bath and extra kettle.
5. Make a 30% or 40% syrup, depending on your taste.
6. Heat the grapes in the 30% or 40% syrup just until boiling.
7. Pack them into ½-pint or 1-pint jars to within ½ inch of top. Reserve the liquid, cover and keep hot.
8. Pour in the hot syrup to within ½ inch of top.
9. Run a nonmetal spatula down the side of each jar to release air bubbles.
10. Add additional syrup, if necessary, to fill to within ½ inch of top of jar.
11. Wipe the tops and threads of the jars with a damp cloth.
12. Put on the lids and screw bands as the manufacturer directs.
13. Following Basic Steps for Canning Fruit 12 and 13, process in a boiling water bath — 15 minutes for both ½-pints and pints.
14. Follow Basic Steps for Canning Fruit 14 through 19.

NECTARINES (follow the recipe for Apricots)

PEACHES

Peaches can be canned in a hot or cold pack. Use 30% or 40% syrup or, if the fruit is very juicy, use sugar and follow the hot pack method. You will need all of the Basic Equipment.

2 to 3 pounds = 1 quart, canned
1 lug (22 pounds) = 8 to 12 quarts, canned
1 bushel (48 pounds) = 18 to 24 quarts, canned

1. Choose ripe but firm peaches.
2. Organize and prepare the equipment and work area.
3. Wash the peaches well. Drain and sort by size.
4. Dip the peaches in boiling water ½ to 1 minute to loosen the skins.
5. Dip them in cold water; drain. Peel.
6. Halve the peaches and remove the pits. Drop the halves into salt-vinegar water (2 tablespoons each salt and vinegar to 1 gallon water) until there are enough halves to fill a load of jars in your water bath canner.
7. Start boiling the water in your water bath canner and extra kettle.
8. Prepare a 30% or 40% syrup.
9. Drain and rinse the peach halves.
10. Cold Pack: Pack the peaches, cavity-side down in overlapping layers, in hot pint or quart jars to within ½ inch of the top of each jar. Ladle or pour in the boiling syrup to within ½ inch of top of the jar.
    Hot Pack: Heat the peaches thoroughly in hot 30% or 40% syrup. If the peaches are very juicy, heat them with sugar (about ½ cup sugar for each quart of peaches) just until the sugar dissolves and the peaches are hot. Pack them into hot pint or quart jars to within ½ inch of top. If you run out of syrup, use boiling water to fill the jars to within ½ inch of the top.
11. Run a nonmetal spatula down the side of each jar to release air bubbles.
12. Add additional syrup, if necessary, to fill to within ½ inch of the top of the jar.
13. Wipe the tops and threads of jars with a damp cloth.
14. Put on the lids and screw bands as the manufacturer directs.
15. Follow Basic Steps for Canning Fruit 12 and 13. Process in a boiling water bath. For the cold pack method: 25 minutes for pints, 30 minutes for quarts. For the hot pack method: 20 minutes for pints, 25 minutes for quarts.
16. Follow Basic Steps for Canning Fruits 14 through 19.

PEARS

Use a 30% or 40% syrup, depending on the sweetness of the pears. You will need all the Basic Equipment. Pears can be canned using either the cold or hot pack method.

2 to 3 pounds = 1 quart, canned
1 box (35 pounds) = 14 to 17 quarts, canned
1 bushel (50 pounds) = 20 to 25 quarts, canned

1. Choose ripe but firm pears.
2. Organize and prepare the equipment and work area.
3. Wash the pears well. Drain and sort by size.
4. Peel, halve and core the pears. Drop the halves into salt-vinegar water (2 tablespoons each salt and vinegar to 1 gallon water) until there are enough halves to fill a load of jars in your water bath canner.
5. Follow Steps 7 through 16 of the peach recipe.

PINEAPPLE

Use a 30% syrup or pineapple juice for canning pineapple. You will need all the Basic Equipment.

2 medium pineapples = 1 quart, canned

1. Choose firm, but ripe perfect pineapples.
2. Organize and prepare the equipment and the work areas.
3. Slice, peel, core the pineapple and remove the eyes. Cut the slices into chunks or dice.
4. Start heating the water in the water bath canner; start boiling extra water in a teakettle.
5. Prepare a 30% syrup, or heat pineapple juice.
6. Cold Pack: Pack chunks or pieces into hot pints or quarts to within ½ inch of the top. Pour in boiling syrup or juice to within ½ inch of the top.
   Hot Pack: Simmer chunks or slices in just enough syrup or juice to cover. Simmer until tender, about 10 minutes. Pack into hot pint or quart jars to within ½ inch of top. Pour in boiling syrup to within ½ inch of the top.

7. Run a nonmetal spatula down the side of each jar to release air bubbles.
8. Add additional syrup, if necessary, to fill to within ½ inch of the top of the jar.
9. Wipe the tops and threads of the jars with a damp cloth.
10. Put on the lids and screw bands as the manufacturer directs.
11. Follow Basic Steps for Canning Fruit 12 and 13. Process in a boiling water bath. For the cold pack method: 30 minutes for pints and quarts. For the hot pack method: 20 minutes for pints and quarts.
12. Follow Basic Steps for Canning Fruit 14 through 19.

PLUMS

Green gage or other very meaty plums are best for canning. Juicy plums tend to fall apart. Use 40% or 50% syrup and all the Basic Equipment, plus a needle.

1½ to 2½ pounds = 1 quart, canned
1 lug (24 pounds) = 12 quarts, canned
1 bushel (56 pounds) = 24 to 30 quarts, canned

1. Choose firm but ripe plums that are in perfect condition.
2. Organize and prepare the equipment and work area.
3. Wash plums well, drain and sort by size.
4. Prick the plum skins with a needle to prevent bursting.
5. Start heating water in the water bath canner; start boiling extra water in a teakettle.
6. Prepare a 40% or 50% syrup, depending on your taste.
7. Cold Pack: Pack plums into hot pints or quarts to within ½ inch of top. Pour in boiling syrup to within ½ inch of top of each jar.
   Hot Pack: Heat the plums in boiling syrup for 2 minutes. Pack into hot pints or quarts to within ½ inch of the top. Pour in boiling syrup to within ½ inch of top of jar.
8. Wipe the tops and threads of the jars with a damp cloth.
9. Put on the lids and screw bands as the manufacturer directs.
10. Follow Basic Steps for Canning Fruit 12 and 13. Process in a boiling water bath — 20 minutes for pints, 25 minutes for quarts.
11. Follow Basic Steps for Canning Fruit 14 through 19.

RHUBARB

Rhubarb can be processed with the cold pack method using a 40% to 50% syrup. Use sugar if you want to can rhubarb using the hot pack method.

2 pounds = about 1 quart, canned

1. Choose tender, nicely colored stalks. Discard the leaves.
2. Organize and prepare the equipment and the work area.
3. Wash the rhubarb well, trim off the ends, but do not peel.
4. Cut the stalks into 1-inch lengths.
5. Start heating the water in the water bath canner; start boiling extra water in a teakettle.
6. Cold Pack: Prepare a 40% to 50% syrup. Pack the rhubarb into hot pints or jars to within ½ inch of top. Pour in the boiling syrup to within ½ inch of top of jar.
   Hot Pack: Mix ½ to 1 cup sugar with each quart of cut rhubarb. Let it stand an hour or so to draw out the juices. Heat the rhubarb and any juice to boiling. Pack it into hot pints or jars to within ½ inch of top. If you run out of syrup, use boiling water to fill the jars to within ½ inch of top.
7. Run a nonmetal spatula down the side of each jar to release air bubbles.
8. Add additional syrup, if necessary, to fill to within ½ inch of top of jar.
9. Wipe the tops and threads of the jars with a damp cloth.
10. Put on the lids and screw bands as the manufacturer directs.
11. Follow Basic Steps for Canning Fruit 12 and 13. Process in a boiling water bath — 15 minutes for cold pack pints or quarts, 10 minutes for hot pack pints or quarts.
12. Follow Basic Steps for Canning Fruit 14 through 19.

STRAWBERRIES

Strawberries simply do not can well — they lose flavor, color and texture. Freeze them instead of trying to can them.

Canning Vegetables

Did your pumpkin plants produce a bumper crop? Canning may be a better idea than serving your family pumpkin for breakfast, lunch and dinner. Canning is one way of preserving high quality fresh supplies of pumpkin, peas, corn and other vegetables. The fresher the vegetable, the tastier (and safer) it will be when canned. In fact, produce rushed from your own garden to your pressure canner is the best for canning. Most vegetables may be cold or hot packed.

All vegetables must be processed in a pressure canner at 10 pounds pressure. If you live above sea level, you need more pressure for food to reach 240°F, so add 1 pound pressure for each 2000 feet above sea level. Thus, if you live at 2000 feet, process at 11 pounds pressure; at 4000 feet, 12 pounds pressure, and so on. If your pressure canner has a weighted gauge, follow manufacturer's directions.

Never skimp on processing times. Even the most perfect vegetables can be spoiled if the processing is not long enough. Always be sure that the heat under the canner is maintained, so that the pressure does not vary. The instruction book that comes with your canner will give you full instructions. Read it carefully before you can for the first time; brush up on the instructions before each canning season.

Basic Equipment

The basic equipment for canning vegetables includes all the Basic Equipment for Canning, except the boiling water bath canner. The difference between canning most fruit and canning vegetables is that vegetables must be processed in a pressure canner. See Basic Equipment for Canning for a detailed description of the following equipment.

Pressure canner
Standard Jars (1 pint, 1½ pint, 1 quart) with 2-piece self-sealing lids
Preserving kettle (for the hot pack method)
Teakettle
Strainer
Spoons — wooden and slotted
Knives — for paring and chopping
Measuring spoons
Wide-mouth funnel and ladle
Jar lifter or tongs
Timer
Hot pads
Nonmetal spatula, plastic knife
Wire cooling racks or folded dish towels

Basic Ingredients

Fresh, perfect, uniformly sized vegetables should be selected for canning. That means you will have to spend time picking over the vegetables, discarding poor-quality produce and sorting canning-quality vegetables by size. For many vegetables, uniform size is important for even, thorough processing. Asparagus, green (snap) and wax beans, carrots, lima beans, beets, corn, greens, okra, peas and summer squash should all be tender, young, just ripe and as freshly picked as possible. Pumpkins, winter squash and sweet potatoes should be plump and unshriveled.

Water and salt, or a salt-sugar mix, are the only other ingredients used in the recipes for canning vegetables.

Seasoning

Our recipes call for salt to be added to vegetables before processing. But, you do not need to add the salt unless you wish. It is not a crucial ingredient in canning, just a flavoring addition. The flavors of peas, beets and corn often benefit from the addition of just a little sugar. Mix 1 part salt with 2 parts sugar and add 2 teaspoons of the salt-sugar mixture to each pint jar before sealing. Double the amount for quarts.

Basic Steps

Always follow the directions that come with your canner. The steps that follow are to show you the entire sequence of vegetable canning, but you must follow the manufacturer's instructions for heating, venting, and generally operating a canner.

1. Select perfect, just mature and very fresh vegetables that are free from blemishes or decay. Sort them by size and maturity and handle the ones that are alike together.
2. Set out all the ingredients and equipment. Wash and dry all equipment, countertops, working surfaces and hands. Wash and rinse the jars, then keep them in hot water, a low oven or a dishwasher on its dry cycle. Prepare the lids as the manufacturer directs.
3. Wash the vegetables very carefully using several changes of washing and rinsing water and scrubbing

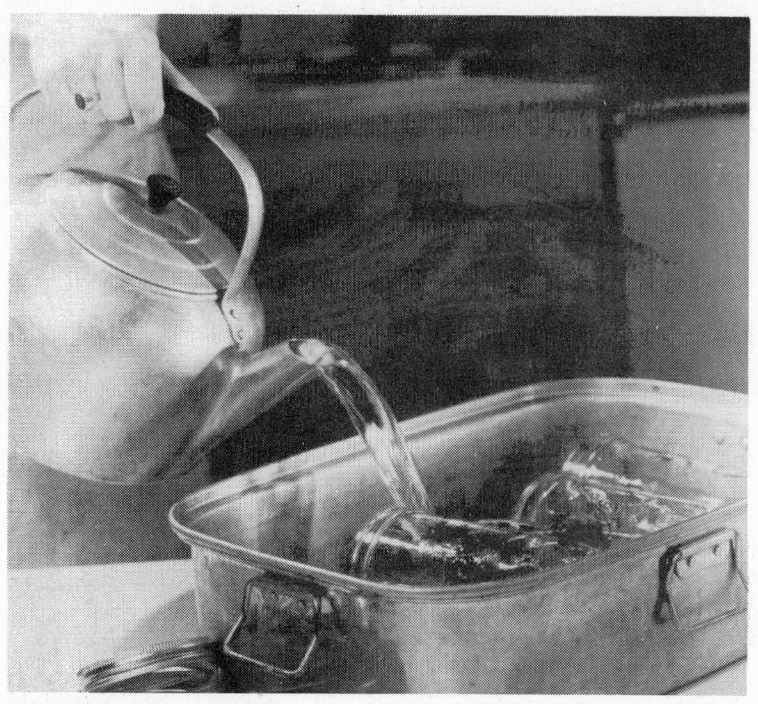

Step 2: Wash and rinse the jars. Keep them hot in a low oven, in hot water or in a dishwasher on its dry cycle.

Step 3: Washing vegetables thoroughly is important to eliminate botulinus bacteria. Use several changes of washing and rinsing water.

Step 4: Prepare the vegetables as the recipe directs. Cut only enough for one canner load at a time.

Step 5: Pack into jars. These green beans are being cold packed, no pre-cooking, to within 1/2 inch of the jar's top.

Step 6: Cold Pack as shown: fill with boiling water to level given in recipe. Hot Pack: pour boiling cooking water into packed jars.

Step 7: Release air bubbles by running slim, nonmetal tool down the sides of the jars. Add more boiling water, if necessary.

them with a brush. Remember that botulinus bacteria are in the soil, and only thorough washing will eliminate them from the vegetables. Be sure that you lift the vegetables out of the rinse water, then let the water drain off.

4. Prepare the vegetables as each recipe directs, cutting, peeling or precooking only enough for one canner load of jars at a time.
5. Pack them into the jars as the recipe directs — this is usually to within ½ inch of the jar's top, but some vegetables expand during processing and need more space.
6. Cold Pack: pour boiling water into the packed jars to the level given in the recipe.
   Hot Pack: pour boiling cooking water into the packed jars to the level given in the recipe.
7. Run a nonmetal spatula, handle of a wooden spoon or other slim tool down the inside of each jar to release any air bubbles. Pour in additional boiling water, if necessary, to bring the liquid back to the level specified in the recipe.
8. Wipe the tops and threads of the jars with a damp cloth.
9. Put on lids and screw bands as the manufacturer directs.

Step 9: After wiping the tops and threads with a damp cloth, put on the prepared lids as the manufacturer directs.

Step 9: Put on screw bands as the manufacturer directs. Protect your hands with a hot pad of some kind.

Step 11: Carefully lower filled, sealed jars into the pressure canner. Arrange jars so steam can flow around them.

Steps 15 and 16: When processing time is up, let canner stand until pressure is reduced. Open by lifting cover away from you.

10. Put the rack in the canner, pour in the boiling water as the manufacturer directs.
11. Carefully lower the filled and sealed jars into the canner, arranging them on the rack so steam can flow around the jars.
12. Put on the cover, gauge and lock following the manufacturer's directions.
13. Heat, following the manufacturer's directions for the steam flow, and the time to exhaust the canner. Put on the control or close the vent.
14. When the canner reaches the required pressure (usually 10 pounds of pressure, adjusting to higher altitudes, if needed), start timing for the exact length of time given in each recipe.
15. When the processing time is up, remove the canner from the heating element to the range top and let it stand until the pressure is reduced. Do not try to hurry this step — it is very important for the pressure to go down slowly. A canner with a weighted gauge may take up to 45 minutes. Nudge the control with a pencil and if you do not see any

steam, the pressure is down. A dial gauge will show the pressure at zero when the jars are ready to be removed.

16. Remove the weight control (if you have that type of canner) and unlock the cover. Open by lifting the cover away from you, so the steam will come out on the far side.
17. Using long handled tongs or a special jar lifter, carefully lift out the jars and put them several inches apart on a folded towel or rack in an out-of-the way, draft-free spot.
18. Let the jars cool completely, 12 to 24 hours. Do not cover the jars while they cool.

Step 17: Use long handled tongs or special jar lifters to take jars out of the canner. Cool on rack or folded towel.

19. When the jars are completely cooled, check the seals. The lids should be slightly depressed and, when the jar is tipped slightly, there should be no leaks. If the center of the lid can be pushed down and springs back up, the canning process did not work. Use the food immediately (store it in the refrigerator) or pour the food into another clean, hot jar, seal with new lid, put a screw band on and reprocess.

20. Wipe the jars with a clean damp cloth, then label clearly with product, date and batch (if more than one canner

Step 19: Check seals when jars are cool. Lids should be slightly depressed and there should be no leaks when tipped.

Steps 20 and 21: Wipe jars clean, then label clearly with product date and batch. Remove screw bands after 36 hours.

Step 23: Check for signs of spoilage before using: bulging lids, mold, broken seals, leakage, spurting, slimy food or bad smell.

Step 23: Open jar with pointed end of opener to pierce the lid so you will not accidently use it again.

load is done in one day).

21. After 36 hours, remove the screw bands. If they are left on they may rust in place. To remove stuck screw bands, wring out a cloth in hot water, then wrap around the band for a minute or two to help loosen it. Clean the bands and store them in a dry place.

22. Store the jars in a cool, dark, dry place where they will not freeze.

23. Before using canned food, check for signs of spoilage. If you notice bulging lids, broken seals, leakage, spurting

Step 24: Heat home-canned vegetables to boiling, cover and boil 15 minutes as a further precaution before tasting.

liquid, mold, a bad smell or food that looks slimy, DO NOT USE IT. Discard the food where humans and animals cannot find it. You can salvage the jars. Wash them thoroughly, rinse and then boil for 15 minutes.

24. As a further precaution, heat home-canned foods to boiling, cover and boil 15 minutes. If the food foams, looks spoiled or smells bad when heated, get rid of it.

If you have followed directions, and if your pressure canner is in good working order, home-canned foods should be safe. We recommend boiling as extra insurance.

Recipes

ASPARAGUS

Home-canned asparagus is handy for cold salads as well as for hot vegetable dishes. You will need all the Basic Equipment.

2½ to 4½ pounds = 1 quart, canned
1 bushel (45 pounds) = 11 quarts, canned

1. Choose tender, evenly-sized, fresh spears.
2. Organize and prepare the equipment and work area.
3. Wash the asparagus spears thoroughly. Trim off the scales and tough ends and wash them again.
4. Cut the spears into 1-inch lengths.
5. Cold Pack: Gently but firmly pack the asparagus as tightly as possible into hot jars to within ½ inch of top. If you want, add ½ teaspoon salt to pints, 1 teaspoon salt to quarts. Pour in boiling water to within ½ inch of top of jar. Hot Pack: Cook the asparagus pieces 2 or 3 minutes in enough boiling water to cover them. Pack the pieces loosely into hot pint or quart jars to within ½ inch of the top. Add salt, ½ teaspoon salt to pints, 1 teaspoon salt to quarts, if you want. Pour in boiling water if you do not have enough cooking water for all the jars.
6. Run a nonmetal spatula down the side of each jar to release air bubbles. Add more boiling liquid to fill to within ½ inch of the top.
7. Wipe the tops and threads of the jars with a damp cloth.
8. Put on the lids and screw bands as the manufacturer directs.
9. Process at 10 pounds of pressure 25 minutes for pints, 30 minutes for quarts. Follow the manufacturer's directions for your canner.
10. Follow Basic Steps for Canning Vegetables 11 through 24.

BEANS, GREEN AND WAX

Even a small plot of beans can produce a surprisingly large crop. Canned, you can enjoy your garden all year. All the Basic Equipment is used.

1½ to 2½ pounds = 1 quart, canned
1 bushel (30 pounds) = 15 to 20 quarts, canned

1. Choose young, tender beans that snap easily.
2. Organize and prepare the equipment and work area.
3. Wash the beans very well in several waters.
4. Trim off the ends and cut or break them into 1-inch pieces.
5. Cold Pack: Pack the beans tightly into hot jars, leaving ½ inch at the top. Add ½ teaspoon salt to each jar, if you want. Pour in boiling water to within ½ inch of the top.
   Hot Pack: Cook the beans in boiling water to cover for 5 minutes. Pack into hot jars to within ½ inch of top. Add ½ teaspoon salt to each jar, if you want. Pour in boiling cooking water to within ½ inch of the top. Use boiling water if you run out of cooking water.
6. Run a nonmetal spatula down the side of each jar to release air bubbles. Add more boiling liquid to fill to within ½ inch of the top.
7. Wipe the tops and threads of jars with a damp cloth.
8. Put on the lids and screw bands as the manufacturer directs.
9. Process at 10 pounds of pressure, 20 minutes for pints, 25 minutes for quarts. Follow the manufacturer's directions for your canner.
10. Follow Basic Steps for Canning Vegetables 11 through 24.

BEANS, LIMA

Perhaps you can employ some young helpers to shell and sort the limas. While they shell, you can organize all the Basic Equipment. You will not need a knife.

3 to 5 pounds (in pods) = 1 quart, canned
1 bushel (32 pounds) = 6 to 10 quarts, canned in pods

1. Choose only young, tender beans.
2. Organize and prepare the equipment and work area.
3. Shell and wash the beans well. Sort them according to size.
4. Cold Pack: Pack small beans loosely into hot jars to within 1 inch of top of pint jars, 1½ inches of top of quarts. Larger beans need ¾ inch headspace in pint jars, 1¼ inches in quarts. Do not press down or shake beans to pack more in the jar. They need space to expand as they cook. If you want, add ½ teaspoon salt to pint jars, 1 teaspoon to quarts. Pour in boiling water to within ½ inch of the top of the jar.
   Hot Pack: Cook the beans in just enough boiling water to cover. Cook only until boiling. Pack the beans into hot jars to within 1 inch of top. Add ½ teaspoon salt to pints, 1 teaspoon to quarts, if you want. Pour in boiling water to within 1 inch of the top.
5. Run a nonmetal spatula down the side of each jar to release air bubbles. Add more boiling liquid to fill within ½ inch of the top.
6. Wipe the tops and threads of jars with a damp cloth.
7. Put on the lids and screw bands as the manufacturer directs.
8. Process at 10 pounds pressure, 40 minutes for pints, 50 minutes for quarts. Follow the manufacturer's directions for your canner.
9. Follow Basic Steps for Canning Vegetables 11 through 24.

BEETS

Beets have to be cooked before they are processed, so be sure to have a large kettle ready along with the rest of the Basic Equipment.

2 to 3½ pounds (without tops) = 1 quart, canned
1 bushel (52 pounds with tops) = 15 to 24 quarts, canned

1. Choose young, tender beets.
2. Organize and prepare the equipment and work area.
3. Sort by size. Cut off the tops, leaving the root and 1 inch of stem.
4. Wash very well.
5. Put the beets in a large saucepan or kettle, cover with boiling water and cook until tender, about 15 to 25 minutes.
6. Drain, slip off the skins and trim the ends and root.
7. Leave tiny beets whole, slice or cube medium or large beets.
8. Pack the beets into hot pint or quart jars to within ½ inch of the top.
9. Add ½ teaspoon salt to pints, 1 teaspoon to quarts, if you choose.
10. Pour in boiling cooking water to within ½ inch of top. Run a nonmetal spatula down the side of each jar to release air bubbles. Add more boiling liquid to within ½ inch of the top.
11. Wipe the tops and threads of the jars with a damp cloth.
12. Put on the lids and screw bands as the manufacturer directs.
13. Process at 10 pounds of pressure, 30 minutes for pints, 35 minutes for quarts. Follow the manufacturer's directions for your canner.
14. Follow Basic Steps for Canning Vegetables 11 through 24.

CARROTS

Large, mature carrots do not can well. You will use all the Basic Equipment for canning young, tender carrots.

2 to 3 pounds (without tops) = 1 quart, canned
1 bushel (50 pounds without tops) = 16 to 25 quarts, canned

1. Choose only young, tender carrots.
2. Organize and prepare the equipment and work area.
3. Wash well, scrape or peel and cut off tops and tips.
4. Slice or dice the carrots.
5. Cold Pack: Pack carrots tightly into hot pint or quart jars to within 1 inch of top. If you want add ½ teaspoon salt to pints, 1 teaspoon to quarts. Pour in boiling water to within ½ inch of top.
   Hot Pack: Put the carrots in a large saucepan, cover with boiling water and heat to boiling. Pack them into hot pint or quart jars to within ½ inch of top. Add ½ teaspoon salt to pints, 1 teaspoon to quarts if you want. Pour in boiling cooking water to within ½ inch of top. Add boiling water if you run out of cooking water.
6. Run a nonmetal spatula down the side of each jar to release air bubbles. Add more boiling liquid to within ½ inch of the top.
7. Wipe the tops and threads of the jars with a damp cloth.
8. Put on the lids and screw bands as the manufacturer directs.
9. Process at 10 pounds of pressure, 25 minutes for pints, 30 minutes for quarts. Follow the manufacturer's directions for your canner.
10. Follow Basic Steps for Canning Vegetables 11 through 24.

CORN, CREAM-STYLE

You do not add real cream to the kernels; the cream is juice from the cob which you scrape off with the back of a knife. You will need all the Basic Equipment. If you grow your own corn, pick it in small quantities (2 to 3 dozen ears at a time) and rush it to the kitchen. This recipe calls for pint jars.

3 to 6 pounds corn (in husks) = 2 pints, canned
1 bushel (35 pounds in husks) = about 12 to 20 pints, canned

1. Home-grown corn, or very fresh corn is best. Work with small quantities, 2 or 3 dozen ears at a time.
2. Organize and prepare the equipment and work area.
3. Husk corn and remove all silk. Wash it well.
4. Cut the corn from cob at about center of the kernel.
5. Scrape the cobs with back of a knife to remove the "cream." Mix the cream with the corn.
6. Cold Pack: Pack corn into hot jars only to within 1½ inches of top. Do not shake or press down; the corn needs room to expand as it cooks. Add ½ teaspoon salt to each jar, if you want. Pour in boiling water to within ½ inch of the top.
   Hot Pack: Measure the corn into a large saucepan or pot and add 1 pint boiling water for each quart of corn. Heat to boiling. Pack the corn and liquid into hot pint jars only to within 1 inch of the top. If you want add ½ teaspoon salt to each jar.
7. Run a nonmetal spatula down the side of each jar to release air bubbles. Add more boiling liquid to within ½ inch of the top.
8. Wipe the tops and threads of the jars.
9. Put on the lids and screw bands as the manufacturer directs.
10. Process at 10 pounds of pressure, 95 minutes for cold pack pints, 85 minutes for hot pack pints. Follow the manufacturer's directions for your canner.
11. Follow Basic Steps for Canning Vegetables 11 through 24.

CORN WHOLE-KERNEL

For this kind of corn, you cut the kernel off closer to the cob than for cream-style corn. Use fresh corn and work with 2 or 3 dozen ears at a time. All the Basic Equipment is used.

3 to 6 pounds corn (in husks) = 1 quart, canned
1 bushel (35 pounds in husks) = 6 to 10 quarts, canned

1. Choose corn as fresh as possible, preferably picked from your own garden and rushed to your kitchen.
2. Organize and prepare the equipment and work area.
3. Husk corn and remove silk. Wash well.
4. Cut the corn from cob at about 2/3 the depth of the kernel. Do not scrape cob.
5. Cold Pack: Pack the corn loosely into hot pint or quart jars to within 1 inch of top. Do not shake or press down; the corn needs room to expand as it cooks. Add ½ teaspoon salt to pints, 1 teaspoon to quarts, if you choose. Pour in boiling water to within ½ inch of top.
   Hot Pack: Measure corn into large saucepan or pot and add 1 pint boiling water for each quart of corn. Heat to boiling. Pack the corn and liquid into hot pint or quart jars to within 1 inch of top, being sure that the corn is covered with cooking water. If you want, add ½ teaspoon salt to pints, 1 teaspoon to quarts.
6. Run a nonmetal spatula down the side of each jar to release air bubbles. Add more boiling liquid to within ½ inch of the top.
7. Wipe the tops and threads of the jars with a damp cloth.
8. Put on lids and screws bands as the manufacturer directs.
9. Process at 10 pounds of pressure, 55 minutes for pints, 85 minutes for quarts. Follow the manufacturer's directions for your canner.
10. Follow Basic Steps for Canning Vegetables 11 through 24.

GREENS (including Spinach)

Collard greens, mustard greens, chard or spinach can be canned with this recipe. You will need all the Basic Equipment plus a vegetable steamer made with a rack in the bottom of a large saucepan. Put the greens in a cheesecloth bag when you steam them. This recipe is for the hot pack method only.

2 to 6 pounds = 1 quart, canned
1 bushel (18 pounds) = 3 to 8 quarts, canned

1. Choose freshly picked, young and tender greens.
2. Organize and prepare the equipment and the work area.
3. Wash greens thoroughly in several waters and pick over carefully.
4. Cut out any tough stems and mid-ribs.
5. Steam the greens by putting a rack in the bottom of a large saucepan or kettle. Pour in just enough water to cover the bottom of the pan but not touch the rack. Heat to boiling. Put about 2½ pounds greens in a cheesecloth bag, place the bag on the rack above the boiling water. Cover and steam 10 minutes or until very wilted.
6. Pack cooked greens very loosely into hot pint or quart jars to within ½ inch of top. Add ½ teaspoon salt to each jar, if you want.
7. Pour in boiling water to within ½ inch of the top. Run a nonmetal spatula down the side of each jar to release air bubbles. Add more boiling liquid to within ½ inch of the top.
8. Wipe the tops and threads of jars with a damp cloth.
9. Put on the lids and bands as the manufacturer directs.
10. Process at 10 pounds pressure, 70 minutes for pints, 90 minutes for quarts. Follow the manufacturer's directions for your canner.
11. Follow Basic Steps for Canning Vegetables 11 through 24.

OKRA

Okra has to be cooked briefly before processing, so have a saucepan or kettle ready in addition to all the Basic Equipment.

1½ pounds = 1 quart, canned
1 bushel (26 pounds) = 16 to 18 quarts, canned

1. Choose small, young okra — freshly picked if possible.
2. Organize and prepare the equipment at the work area.
3. Wash okra well and trim off the stem ends.
4. Cook in boiling water for 1 minute.
5. Slice or leave them whole.
6. Pack them into hot pint or quart jars to within ½ inch of the top.
7. Add ¼ teaspoon salt to pints, ½ teaspoon to quarts, if you want.
8. Pour in boiling cooking water to within ½ inch of top. Use boiling water, if you run out of cooking water. Run a non-metal spatula down the side of each jar to release air bubbles. Add more boiling liquid to within ½ inch of the top.
9. Wipe off the tops and threads of the jars.
10. Put on the lids and screw bands as the manufacturer directs.
11. Process at 10 pounds pressure, 25 minutes for pints, 40 minutes for quarts.

PEAS, GREEN

Add a sugar-salt mixture to bring out the flavor, if you want. Mix 1 part salt with 2 parts sugar and add 2 teaspoons of the salt-sugar mixture to each pint jar before sealing. You will need all the Basic Equipment.

(Continued On Next Page)

3 to 6 pounds in pods = *1 quart, canned*
1 bushel (30 pounds) in pods = *10 to 12 quarts, canned*

1. Choose young, tender peas.
2. Organize, and prepare the equipment and the work area.
3. Shell, then wash the peas.
4. Cold Pack: Pack the peas loosely into hot pint or quart jars to within 1 inch of the top. Do not shake or press down; the peas need room to expand as they cook. Add ½ teaspoon salt to pints, 1 teaspoon to quarts, if you like. Pour in boiling water to within 1½ inches of the top.
   Hot Pack: Put the peas in a large saucepan, cover with boiling water and heat to boiling. Pack loosely into hot pint or quart jars to within 1 inch of top. Add ½ teaspoon salt to pints, 1 teaspoon to quarts, if you want. Pour in boiling cooking water to within 1 inch of top.
5. Run a nonmetal spatula down the sides of each jar to release air bubbles. Add more boiling liquid to within 1 inch of the top for hot pack, 1½ inches of the top for cold pack.
6. Wipe the tops and threads of the jars.
7. Put on the lids and screw bands as the manufacturer directs.
8. Process at 10 pounds of pressure, 40 minutes for quarts and pints. Follow the manufacturer's directions for your canner.
9. Follow Basic Steps for Canning Vegetables 11 through 24.

PUMPKIN

Pumpkin is rather mild tasting when prepared as a vegetable rather than a pie. You can save the seeds and toast them. You will need all the Basic Equipment. Pumpkin has to be cooked before processing it. This recipe gives you a choice of cubed or strained pumpkin. Strained pumpkin requires a steamer and sieve or food mill.

1½ to 3 pounds = 1 quart, canned

1. Choose a fresh, unshriveled pumpkin.
2. Organize and prepare the equipment and the work area.

3. Wash the pumpkin, cut and remove the seeds. Pare.
4. Cut it into 1-inch cubes.
5. Cubed: Put the pumpkin pieces in a large saucepan, add boiling water to cover and heat to boiling. Pack the cubes into hot jars to within ½ inch of top. Add ½ teaspoon salt to pints, 1 teaspoon to quarts, if you like. Pour in boiling cooking water to within ½ inch of the top. Run a nonmetal spatula down the side of each jar. Add more boiling liquid to fill to within ½ inch of the top.
   Strained: Put a rack in the bottom of a large saucepan or kettle. Pour in just enough boiling water to cover the bottom of the pan without touching the rack; heat to boiling. Arrange the pumpkin pieces on the rack and steam 25 minutes or until tender. Press them through sieve or food mill. Put the strained pumpkin in a pan and simmer until thoroughly hot. Pack into hot pint or quart jars to within ½ inch of the top.
6. Wipe the tops and threads of the jars with a damp cloth.
7. Put on the lids and screw bands as the manufacturer directs.
8. Process at 10 pounds of pressure. For cubed pumpkin: 55 minutes for pints, 90 minutes for quarts. For strained pumpkin: 65 minutes for pints, 80 minutes for quarts. Follow the manufacturer's directions for your canner.
9. Follow Basic Steps for Canning Vegetables 11 through 24.

SUMMER SQUASH

Summer squash (yellow, crook neck, patty pan or zucchini) is another home garden plant that sometimes overwhelms you with its production. You will need all the Basic Equipment to can summer squash.

2 to 4 pounds = 1 quart, canned
1 bushel (40 pounds) = 10 to 20 quarts, canned

1. Choose young, tender, thin-skinned squash.
2. Organize and prepare the equipment and work area.
3. Wash well, but do not peel. Trim off ends.
4. Cut the squash into ½-inch slices or chunks (if large).
5. Cold Pack: Pack squash tightly into hot pint or quart jars to within 1 inch of top. If desired, add ½ teaspoon salt to pints, 1 teaspoon to quarts. Pour in boiling water to within ½ inch of the top.
   Hot Pack: Put squash pieces in a large saucepan, add boiling water to cover and heat to boiling. Pack it loosely into hot pint or quart jars to within ½ inch of the top. Add ½ teaspoon salt to pints, 1 teaspoon to quarts, if you like. Pour in boiling cooking water to within ½ inch of top. Use boiling water if there is not enough cooking water.
6. Run a nonmetal spatula down the side of each jar to release air bubbles. Add more boiling liquid to within ½ inch of the top.
7. Wipe the lids and screw bands as the manufacturer directs.
9. Process at 10 pounds of pressure. For the cold pack method: 25 minutes for pints, 30 minutes for quarts. For the hot pack method: 30 minutes for pints, 40 minutes for quarts. Follow the manufacturer's directions for your canner.
10. Follow Basic Steps for Canning Vegetables 11 through 24.

WINTER SQUASH (follow the recipe for Pumpkin)

SOYBEANS (shelled, green)

You must really rush soybeans from field to jar — they start to lose flavor almost as soon as they are picked. The beans have to be briefly boiled before processing. All the Basic Equipment is used. You will need an extra kettle and ice water to soften the pods.

4 to 5 pounds beans in pods = 1 quart, canned

1. Use fresh picked soy beans.
2. Organize and prepare the equipment and work area.
3. Wash the pods of the beans well.
4. Put the pods in a large saucepan or kettle, cover with boiling water and boil 5 minutes.
5. Cool the pods immediately in ice water.
6. Snap the pods in two and squeeze out the beans. Wash the beans.
7. Put the washed beans in a large saucepan or pot, add boiling water to cover and boil 3 to 4 minutes.
8. Pack into hot pint or quart jars to within 1 inch of the top.
9. Pour in boiling cooking water or boiling water to within 1 inch of the top. Run a nonmetal spatula down the side of each jar to release air bubbles. Add more boiling liquid to within 1 inch of the top.
10. Add ½ tablespoon sugar and 1 teaspoon salt to each jar, if you wish.
11. Wipe the tops and threads of the jars with a damp cloth.
12. Put on the lids and screw bands as the manufacturer directs.
13. Process at 10 pounds pressure, 60 minutes for pints, 70 minutes for quarts. Follow the manufacturer's directions for your canner.
14. Follow Basic Steps for Canning Vegetables 11 through 24.

SWEET POTATOES

Sweet potatoes have to be boiled or steamed before processing. All the Basic Equipment is used. Sweet potatoes can be dry packed with no extra water or wet packed with extra liquid — boiling water or a 40% syrup.

2 to 3 pounds = 1 quart, canned
1 bushel (50 pounds) = 16 to 25 quarts, canned

1. Select plump, unshriveled sweet potatoes.
2. Organize and prepare the equipment and work area.
3. Wash sweet potatoes well and sort according to size.
4. Boil or steam about 20 to 30 minutes, or until tender and skins slip off easily.
5. Peel and cut into slices or pieces.
6. Dry Pack: Pack the slices or pieces tightly into hot pint or quart jars, pressing gently to fill the spaces. Fill to within 1 inch of the top.
   Wet Pack: Pack the slices or pieces into hot jars to within 1 inch of the top. Add ½ teaspoon salt to pints, 1 teaspoon to quarts, if desired. Pour in boiling water or a 40% syrup to within 1 inch of the top. Run a nonmetal spatula down the side of each jar to release air bubbles. Add more boiling liquid to within ½ inch of the top.
7. Wipe the tops and threads of jars with a damp cloth.
8. Put on the lids and screw bands as the manufacturer directs.
9. Process at 10 pounds of pressure. For dry pack: 65 minutes for pints, 95 minutes for quarts. For wet pack: 55 minutes for pints, 90 minutes for quarts. Follow the manufacturer's directions for your canner.
10. Follow Basic Steps for Canning Vegetables 11 through 24.

Canning Tomatoes

Called Love Apples when first introduced to Europe in Elizabethan times, the tomato was thought to be an aphrodisiac and possibly poisonous if eaten raw — a feeling that persisted in some parts of the United States until late in the 19th century. The general suspicion was that anything that gorgeous couldn't be good for you.

In spite of their reputation, Thomas Jefferson planted tomatoes and they held a part in the American diet, though they were usually thoroughly disciplined by long cooking, pickling or preserving. Enlightened men like George Washington Carver in the mid-19th century made public displays of eating quantities of raw tomatoes to demonstrate that tomatoes would not hurt you. The horrified crowds, expecting the foolhardy tomato-eaters to fall frothing on the ground, were surprised to see them survive.

Tomato plants can overwhelm you with their productivity (undoubtedly another sign of their wanton natures — don't forget the slang use of "tomato"). And, sometimes tomato plants have a thoughtless habit of ripening all their fruit at once. When you have more tomatoes than you can eat or give away, canning them naturally comes to mind. But before you start "putting up" your tomatoes, find out if they are high acid or low acid fruit. All the nasty old-fashioned rumors about poisonous tomatoes may come true if low acid tomatoes are improperly canned.

Botulism

Botulism has a hard time developing in high acid foods. Tomatoes used to be dependably high-acid fruits that could be safely processed in a boiling water bath. Now, however, there are new sweet tomatoes that do not have a high enough acid content to prevent botulinus growth. They must have acid added or be processed in a pressure canner. If you intend to grow your own tomatoes for canning, check with your local state or county agricultural extension office for a list of tomato varieties that are high in acid. If you are in doubt about the variety of tomatoes you buy or grow for canning, treat them as though they were low acid tomatoes.

Low Acid Tomatoes

Yellow, pink or white hybrid tomatoes are definitely low in acid. Tomatoes that you know are low in acid, or those you are not sure of, must have acid added to them or must be processed in a pressure canner at 10 pounds of pressure for 10 minutes. If you already own a pressure canner you have no problem at all — just use it for processing tomatoes.

Adding Acid

If you do not own a pressure canner and do not want to invest in one, then you must add acid. To each quart tomatoes or juice, add any one of the following:

- ½ teaspoon crystalline citric acid
- 2 teaspoons bottled, not fresh, lemon juice
- 2 teaspoons vinegar

Citric acid is U.S. Pharmaceutical Crystalline Citric Acid Monohydrate, available from drugstores. Bottled lemon juice has controlled acidity, but fresh lemon juice may vary.

For low acid tomatoes processed in a pressure canner, use the Basic Equipment for Canning Vegetables. Refer to the Basic Steps for Canning Vegetables for details on the safe preparation of the equipment and food. Lids must be prepared as the manufacturer directs. Do not forget to wash, rinse and keep the jars hot, even if you are using the cold pack method.

High Acid Tomatoes

Tomatoes whose high acidity is certain can be processed in a boiling water bath instead of a preserve canner. Recommended times are 45 minutes for cold pack pints and quarts, 30 minutes for hot pack pints and quarts.

For high acid tomatoes, follow the Basic Steps for Canning Fruit and use the Basic Equipment for Canning Fruit. Be sure to wash, rinse and keep the jars hot. The lids, of course, must be prepared as their manufacturer directs.

Seasoning

The salt or sugar-salt mixture called for in the recipes is for flavoring only. Omit if you wish or if you are trying to meet special diet needs. The salt-sugar mixture is a combination of 2 parts sugar to 1 part salt.

Recipes

TOMATOES

Low acid tomatoes have to be processed in a pressure canner or have acid added if they are going to be processed in a boiling water bath. High acid tomatoes can be processed in a boiling water bath. You will need all the Basic Equipment for Canning Vegetables.

2½ to 3½ pounds = 1 quart, canned
1 lug (30 pounds) = 10 quarts, canned
1 bushel (53 pounds) = 15 to 20 quarts, canned

1. Choose fresh, firm, ripe, perfect tomatoes. Make sure they do not have any black spots, cracks or soft spots. Imperfect tomatoes could harbor microorganisms that are harmful, so do not use them for canning.

Step 3: Dip tomatoes in boiling water to loosen their skins. Then dip in cold water, slip off skins.

Step 6: Cold pack cored, quartered tomatoes gently but firmly into hot jars to within 1/2 inch of top.

2. Organize and prepare the equipment and work area.
3. Dip the tomatoes into boiling water for a minute or two to loosen the skins. Then dip them in cold water. Slip off the skins.
4. Cut out the core. Leave them whole or cut them in halves or quarters.
5. If you are using a water bath canner, start heating the water.
6. Cold Pack: Pack the tomatoes gently but firmly into hot pint or quart jars to within ½ inch of the top, pressing to fill space. Add no water. If you want, season with ½ teaspoon salt for pints, 1 teaspoon for quarts, or 2 teaspoons sugar-salt mixture (1 part salt mixed with 2 parts sugar)

Step 6: For both hot and cold packs: add 2 teaspoons salt-sugar mix to each quart, if desired.

Step 6: If using low acid tomatoes or if unsure of acidity, use a pressure canner or add citric acid, lemon juice, or vinegar.

Step 7: Run a wooden spoon handle or other slim, nonmetal tool down the sides to release air bubbles.

for each quart. Add acid if you are unsure of the tomatoes' acidity and you are processing them in boiling water bath. Hot Pack: Quarter the tomatoes into a large pan. Heat them to a boil; stir to prevent sticking. Pack into hot or quart jars to within ½ inch of the top. Season as for cold pack. Add acid, if necessary.

7. Run a nonmetal spatula down side of each jar to release air bubbles. Add hot liquid to both cold and hot pack tomatoes, if needed to fill to within ½ inch of the top.
8. Wipe the tops and threads of the jars with a damp cloth.
9. Put on the lids and screw bands as the manufacturer directs.
10. For low acid tomatoes, process at 10 pounds of pressure 10 minutes for pints or quarts (follow Basic Steps for Canning Vegetables 11 through 24).
    For high acid tomatoes, process in a boiling water bath 45 minutes for cold pack pints or quarts, 30 minutes for hot pack pints or quarts (follow Basic Steps for Canning Fruit 11 through 19).

TOMATO JUICE

This recipe can be used either for low or high acid tomatoes. Use a pressure canner or add acid if you are not sure you have high acid tomatoes. You will need all the Basic Equipment for Canning Vegetables, including a sieve or strainer.

3 to 3½ pounds = 1 quart canned juice
1 bushel (53 pounds) = 12 to 16 quarts canned juice

1. Choose perfect, ripe juicy tomatoes.
2. Organize and prepare the equipment and work area.
3. Peel and core as for canned tomatoes.
4. Simmer in a large saucepan or kettle until soft, stirring often.
5. Press through a fine sieve or strainer and return them to the pan or kettle.
6. Heat the juice just to boiling. (If you are using a water bath canner, start heating the water.)
7. Pour or ladle the juice into hot pint or quart jars to within ½ inch of the top.
8. Add 1 teaspoon salt or 2 teaspoons salt-sugar mixture (1 part salt to 2 parts sugar). Add acid if you are unsure of the tomatoes' acidity.
9. Wipe the tops and threads of the jars with a damp cloth.
10. Put on the lids and screw bands as the manufacturer directs.
11. For low acid tomatoes, process at 10 pounds of pressure 10 minutes for pints or quarts (follow Basic Steps for Canning Vegetables 11 through 24).
    For high acid tomatoes, process in a boiling water bath 10 minutes for pints, 15 minutes for quarts (follow Basic Steps for Canning Fruit 11 through 19).

TOMATO JUICE COCKTAIL

A zippy blend, this cocktail is great for appetizers or aspics. Because bottled lemon juice is added for acid, you can process the juice in a boiling water bath. The recipe makes 2 quarts.

Ingredients

2 *quarts tomato juice (prepare as directed in the recipe for tomato juice)*
3 *tablespoons bottled lemon juice*
1 *tablespoon salt*
2 *teaspoons grated celery*
1 *teaspoon prepared horseradish*
1 *teaspoon onion juice*
Dash Worcestershire sauce

Equipment

Use all the Basic Equipment for Canning Fruit
2 (1-quart) jars with 2-piece self-sealing lids

1. Organize and prepare the equipment and the work area.
2. Combine all the ingredients and heat to boiling.
3. Start heating water in the water bath canner.
4. Pour the juice into 2 hot quart jars to within ½ inch of the top.
5. Wipe the top and threads with a damp cloth.
6. Put on the lids and screw bands as the manufacturer directs.
7. Follow Basic Steps for Canning Fruit 11 and 12. Process in a boiling water bath for 30 minutes.
8. Follow Basic Steps for Canning Fruit 14 through 19.

STEWED TOMATOES

This delicious old-fashioned recipe makes 7 pints. You have to process stewed tomatoes in a pressure canner.

Ingredients

4 quarts peeled, cored, chopped tomatoes
1 cup chopped celery
½ cup chopped onion
¼ cup chopped green pepper
1 tablespoon sugar
2 teaspoons salt

Equipment

Use the Basic Equipment for Canning Vegetables
7 (1-pint) jars with 2-piece self-sealing lids

1. Organize and prepare the Basic Equipment and work area.
2. Combine all the ingredients in large kettle, cover and simmer 10 minutes, stirring occasionally.
3. Pour or ladle the tomatoes into hot pint or quart jars to within ½ inch of the top.
4. Run a nonmetal spatula down the side of each jar to release air bubbles.
5. Wipe the tops and threads of the jars with a damp cloth.
6. Put on the lids and screw bands as the manufacturer directs.
7. Process at 10 pounds pressure, 15 minutes for pints, 20 minutes for quarts. Follow the manufacturer's directions for your canner.
8. Follow the Basic Steps for canning Vegetables 11 through 24.

TOMATO PUREE

This recipe requires a pressure canner; the concentration of the cooked tomatoes lowers their acidity. Do not process tomato puree in a water bath canner. You will need all the Basic Equipment for Canning Vegetables, plus a sieve or food mill and large preserving kettle.

The quantity of canned tomato puree will vary greatly, depending on how long you simmer the tomatoes.

1. Select fresh, ripe, juicy tomatoes.
2. Organize and prepare the equipment and work area.
3. Dip the tomatoes into boiling water for a minute or two to loosen the skins. Then dip them in cold water. Slip off the skins and cut out the cores.
4. Cut them into chunks and place in a large preserving kettle.
5. Cover and cook over low heat until the tomatoes are soft.
6. Uncover and simmer over medium heat, stirring frequently, until very, very soft.
7. Press through a sieve or food mill, then return to kettle and simmer until it is the thickness of catsup.
8. Pour or ladle into hot pint or quart jars to within ½ inch of the top.
9. Wipe the tops and threads of jars with a damp cloth.
10. Put on the lids and screw bands as the manufacturer directs.
11. Process at 10 pounds pressure for 15 minutes. Follow the manufacturer's directions for your canner.
12. Follow Basic Steps for Canning Vegetables 11 through 24.

TOMATO PASTE

Another great way to put away the fruit of the vine, this recipe makes 9 (½-pint) jars. Because of the concentration of tomatoes, a pressure canner must be used.

Ingredients

8 quarts peeled, cored, chopped tomatoes
1½ cups chopped sweet red peppers
2 bay leaves
1 tablespoon salt
1 clove garlic, if desired

Equipment

Use all the Basic Equipment for Canning Vegetables
Sieve or strainer
9 (½ pint) jars with 2-piece self-sealing lids

1. Organize and prepare the equipment and work area.
2. Combine all the ingredients, except the garlic, in large preserving kettle and simmer 1 hour.
3. Press through a sieve or strainer.
4. Return the tomatoe paste to the kettle, add the garlic clove and cook slowly, stirring frequently, about 2½ hours or until thick enough to mound on a spoon.
5. Remove the garlic clove.
6. Pour or ladle the puree into hot pint jars to within ½ inch of the top.
7. Wipe the tops and threads of jars with a damp cloth.
8. Put on the lids and screw bands as the manufacturer directs.
9. Process at 10 pounds of pressure for 25 minutes. Follow manufacturer's directions for your canner.
10. Follow Basic Steps for Canning Vegetables 11 through 24.

Pickles, Relishes, Chutneys And Kraut

Nothing complements a meal better than a pretty dish of your very own pickles, tangy relish or spicy chutney. Once you have made and tasted your own chili sauce or catsup, you will probably want to experiment with your own 57 varieties.

Pickles (relishes, chutneys and sauces) are fruits or vegetables prepared with brine (salt and water) or vinegar, some sugar and spices. The vinegar acts as a preservative, keeping any spoilage organisms from growing. Sealing pickled foods in jars and processing in a boiling water bath helps keep the pickles fresh, crisp and free from mold.

Whole fruits or vegetables, slices or pieces, cooked in vinegar or a vinegar-sugar syrup, can become pickles. Chopped or ground combinations cooked with vinegar, sugar and spices become relishes. Chutneys are highly-spiced fruit and/or vegetable combinations.

The old-fashioned dill pickles and sauerkraut are actually fermented in brine, rather than cooked in vinegar. The brine, plus the sugar from the cucumber or cabbage, plus time, promote a special kind of bacterial action that, over several days or weeks, changes cucumbers to pickles and transforms cabbage to kraut.

Pickles were as much appreciated in 1776 as they are today. Just listen to Thomas Jefferson wax eloquent on the subject: "On a hot day in Virginia, I know of nothing more comforting than a fine spiced pickle, brought up trout-like from the sparkling depths of the aromatic jar below the stairs in Aunt Sally's cellar."

The juicy, pungent cucumber pickle is just one of hundreds of pickled foods. The imaginative American cook has invented myriads of variations on the pickle: pickled pumpkin, pickled cranberry, pickled anything-left-in-the-garden. The recipes in this section reflect many regional tastes and influences: hot spiced relishes and sauces from the South and Southwest, sauerkraut from the Pennsylvania Dutch (Germans) and, in the Blue Ribbon Recipes, mustard pickles from Maine, to name a few.

Basic Equipment for Pickles. Includes a plastic pail for fermenting and boiling water bath.

Basic Equipment

Do not use copper, brass, galvanized or iron pots, pans or utensils for brining pickles or heating pickling solutions. They may darken the brine and the pickles.

1. **A crock** is necessary to hold dill pickles or sauerkraut for fermenting. Our recipes call for a five-gallon crock. You could use two smaller crocks, or a larger one. You can find new crocks in hardware and houseware stores; old ones are often sold at antique shops and auctions. They are not cheap. Be sure that old crocks are not cracked. Cheaper to buy, and almost as easy to use, are one-gallon glass jars. These glass jars are the kinds that hold restaurant-sized mayonnaise and salad dressing. While not as easy to pack and unpack as crocks, these jars are an alternative if you do not want to invest in a crock, or if you want to make a fraction of the recipe.
2. **A boiling water bath canner** is an essential. All pickled products must be processed. Use a boiling water bath canner as described in the Canning chapter, or, if you are using ½-pint jars, follow the suggestions for a small water bath canner in the Jelly chapter. You can use a pressure canner for a water bath if you do not fasten the cover in place. Be sure the petcock is open so steam can escape.
3. **A preserving kettle** (8 to 10 quarts) will handle all the recipes that follow. Some smaller recipes call for a medium or large saucepan. Remember, do not use copper, brass, galvanized or iron pans.
4. **A large mixing bowl,** unchipped enamel, glass or even a pottery container can hold pickles and some relishes for short brining. Again, avoid copper, brass, galvanized or iron — they could cause discoloration.
5. **A wide-mouth funnel and ladle** make it much easier to fill the jars. Be sure your ladle has an insulated handle to protect your hand — you will be working with boiling hot mixtures.
6. **Standard glass canning jars** with two-piece self-sealing lids are the containers required for pickles. Dill pickles and kraut can go in quart jars. Other pickles will fit nicely in pint, 12-ounce or ½-pint jars. Pick the size that meets your needs — couples often prefer ½-pints, big families may want pints. The recipes give the yield in one size of

jar, but you can use another size of jar. Convert total amount to fit in the size container you want to use. See conversion chart at the front of the book if you need help translating ½-pints into 12-ounces. Always check jars to be sure they are in mint condition. Run your finger around the top edge — a crack or nick could prevent a perfect seal; discard imperfect jars and lids. Lids must be new, not last year's and not used.

7. **Measuring tools** include dry (metal) and liquid (glass or plastic) measuring cups, and a set of measuring spoons. In addition to 1-cup dry and liquid measures, we also recommend a 1-quart liquid measuring cup and a 2-cup dry measure for convenience in measuring large amounts. If you are going to make kraut, you will need a kitchen scale.
8. **A colander or sieve** is helpful for draining or straining ingredients. A colander lined with cheesecloth can replace a large sieve or strainer.
9. **Cheesecloth** to make into spice bags should be a part of your pickling kit. Cut a small double square to hold whole spices; tie the top with a string or rubber band.
10. **Spoons** are indispensible. Wooden spoons are excellent for stirring. A slotted spoon is used for separating vegetables from liquid, and a teaspoon is handy for adding or subtracting small amounts as you fill jars.
11. **Knives** should be well-sharpened.
12. **A cutting board** will save your countertops.
13. **A food grinder** will make relishes and chutneys easier than hand chopping.
14. **A timer** will help you keep track of cooking times.
15. **Hot pads, oven mitts, wire cooling racks or folded dish towels** will protect your hands and countertops from hot jars.
16. **Long-handled tongs or special jar lifters** are necessary for safe jar handling.

Basic Ingredients

Just as in all other foods for preservation, there are only a few necessary ingredients, but the quality and exact amounts of each are very important in each recipe. Use produce that is as fresh as possible. Take it from the field to your kitchen and into jars just as rapidly as possible. If you cannot process the

Basic Ingredients for Pickles. Success depends on pure granulated salt, vinegar, spices, soft water.

produce immediately, be sure to keep it refrigerated.

1. Fruits and vegetables should be just barely ripe; they will keep their shape better than if they were fully ripe. Peaches and pears can be a little underripe. Always select cucumber varieties that have been created for pickling. The large salad cucumbers were developed for salads, not for pickles. Look for smaller, less pretty cukes, with pale skins, plenty of bumps and black spines. If in doubt about the variety, check with the produce person or your local extension office. Select evenly shaped and sized fruit and vegetables for even cooking and better looking pickles.
2. Water is an important pickle ingredient, especially for long-brined pickles. Soft water is best. Hard water can cloud the brine or discolor the pickles. If you do not have soft water, boil hard water for 15 minutes, then let it stand overnight. Skim off the scum, then carefully dip out what you need so you will not get any sediment from the bottom.
3. Salt, too, makes a difference. Just any salt will not do. Table salt has special substances added to prevent it from

caking in your shaker, and these materials can cloud brine. Iodized salt can darken brine. Use pure granulated salt, also known as kosher salt, pickling salt, or dairy salt. Most supermarkets stock it with canning supplies.

4. Vinegar is a crucial ingredient for many pickle recipes. Check the label when you shop and be sure to get vinegar marked 4% to 6% acidity. Weaker vinegar will not pickle foods. Use distilled white vinegar for light colored pickles, cider vinegar for darker foods or more interesting flavor. Pickle recipes use a lot of vinegar, so buy a big bottle.
5. Sugar can be brown or white granulated, depending on the lightness or darkness of food to be pickled. Or, if you wish, use half corn syrup or half honey and half sugar.
6. Spices must be fresh. Old spices will make your pickles taste musty. Most of our recipes call for whole spices — they give stronger flavor and do not color the pickles. We suggest you tie the spices in a cheesecloth bag and add them to the kettle during cooking, then remove the bag before packing the pickles into jars. Some cooks like to leave whole spices in the jars for stronger flavor and just for appearance's sake, but spices may darken the pickles.

Basic Steps

Always read a recipe through before you start. Some recipes have slightly different procedures than these basic steps and some take several days to complete.

1. Select perfect, just-ripe evenly sized fruits and vegetables. Do not use any fruit with the slightest indication of mold or spoilage.
2. Set out and wash all equipment, wash all working surfaces, as well as your hands.
3. Wash fruits or vegetables well in plenty of cold water. Discard any cucumbers that float in the wash water. Scrub the fruits and vegetables with a brush, but scrub gently so you do not bruise the food. Lift washed produce out of the water to drain on a rack or drainboard. Do not let the water drain out over the food — all the dirt goes right back on. Do not soak fruit or vegetables. Be sure that all portions of the blossoms are removed from cukes. Blossoms contain an enzyme that can soften the pickles.

Step 3: Wash food well. Scrub gently but firmly. Remove blossom ends of cucumbers.

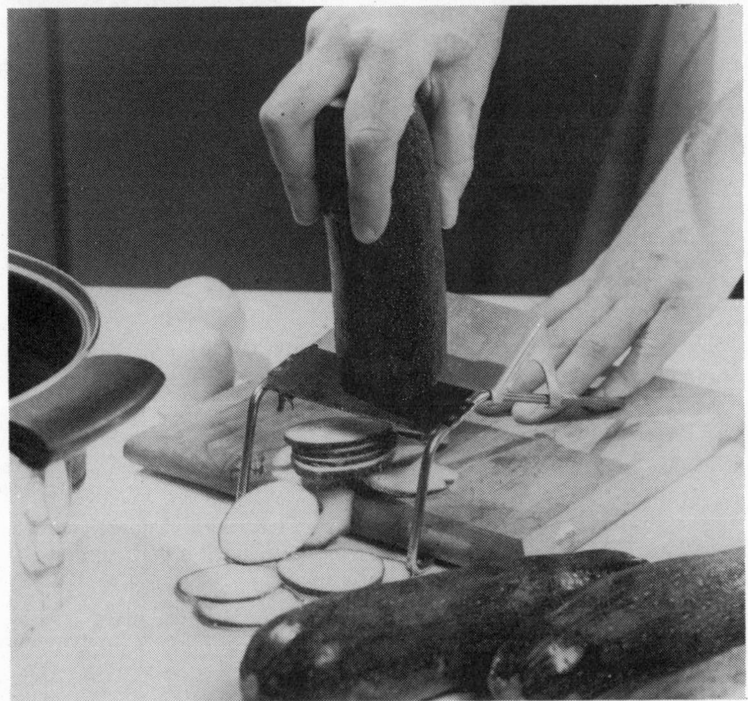

Step 4: Prepare food as recipe directs. These zucchini are being thin-sliced on a stand slicer.

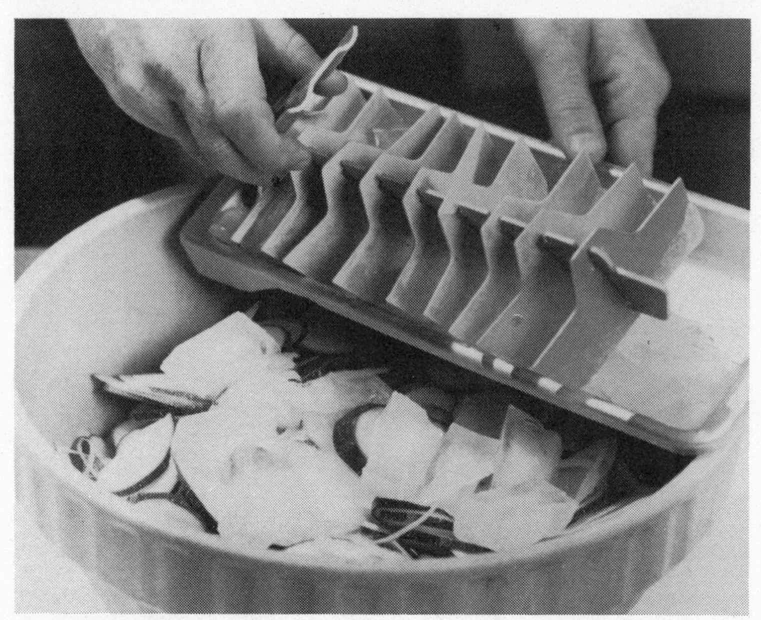

Step 4: Some food requires several hours of preparation. Here, sliced zucchini is covered with ice.

Step 5: Measure and prepare other ingredients. Drained zucchini and onions go into hot vinegar and spices.

Step 8: Pack hot pickles into hot jars, following each recipe's directions. Use a wide-mouth funnel.

Step 8: Ladle hot pickling liquid into jars to within 1/2 inch of the top.

Step 8: Run a spatula handle or other nonmetal, slim tool down sides of jar to release air bubbles.

Step 8: After you wipe the tops and threads with a damp cloth, put on the lids and screw bands as the manufacturer directs.

4. Prepare the produce as the recipe directs. Apples, peaches, pears or other fruits that might discolor can be cut or chopped directly into water with vinegar, salt or an ascorbic acid mixture added. Leave the fruit in this mixture only until the other ingredients are ready, not more than 15 to 20 minutes.
5. Measure and/or prepare all other ingredients.
6. Put water in the water bath canner and start heating it. Wash and rinse the jars; keep them hot. Prepare the lids as the manufacturer directs. If the pickles must stand several hours or overnight, then do this step 15 minutes to one hour before the pickles will be ready to pack into the jars.
7. Prepare the recipe as directed.
8. Pack the pickles into hot jars, following each recipe's directions. Stand the hot jars on a rack or folded towels near the preserving kettle. Put a wide-mouth funnel in the jar and pack. If you notice air bubbles in the jar, gently run a nonmetal spatula or knife down the side of the jar to release the air bubbles. Add additional hot liquid, if needed, to fill to the level given in the recipe. Wipe off the top and threads of the jar with a damp cloth, put on the lid and screw on the band. Set aside and continue to the next jar. Fill and seal only one jar at a time to help prevent spoilage.
9. When all the jars are filled and sealed, carefully lower them, one at a time, into the boiling water bath. When the water returns to a boil, cover and begin timing. (Some recipes tell you to start timing as soon as jars go into boiling water.) Add one minute to the processing time for each 1000 feet above sea level. Thus, if you live at 5000 feet, add five minutes to processing time.
10. When the time is up, carefully lift the jars out of the boiling water and set them several inches apart on a rack or folded towel. Be sure they are out of drafts.
11. Let the jars cool, undisturbed, for 12 to 24 hours.
12. When completely cooled, remove the screw bands and check the seals. Lids should be depressed or, when jar is tipped slightly, there should be no leaks. If the center of the lid can be pushed up and down, or if there are any leaks, use the pickles from the jar immediately because they are improperly sealed. You can pour the pickles into another clean hot jar, seal with new lid, put on the screw band and re-process.
13. Wipe the jars with a clean, damp cloth, then label them

clearly with the product date. Be sure to remove the screw bands. If left on, they could rust in place.

14. Store the jars in a cool, dark, dry place where they will not freeze. Pickles will keep a long time, but you should really only pack enough to last a year.
15. Before using pickles, check for signs of spoilage: bulging lid, leakage, spurting liquid, mold, bad smell, or pickles that are soft, slimy or mushy. If you notice any of these tell-tale signs, do not taste or use the pickles. Discard them where neither people nor animals can find them. You can salvage the jar — wash it well, rinse and then boil it for 15 minutes.
16. Pickles are usually crisper if chilled before serving. Always store opened pickles in the refrigerator.

Nobody's Perfect

There are so many factors involved in pickling — weather and growing conditions, type of salt, acidity of vinegar, storage temperature, time from gathering to pickling, processing — that sometimes things go wrong. Here are some common problems, and causes.

1. **Soft or slippery** pickles could result from not removing the scum from the surface of the brine, or not keeping the cukes submerged in the brine, from using too weak brine or vinegar, using hard water or not removing the blossom of a cucumber, or from storing in too warm a spot.
2. **Shriveled** pickles may be the result of too strong brine, vinegar, syrup or pickling solution, or because cucumbers did not travel to the kitchen fast enough.
3. **Hollow** pickles could result from too long a time between picking and processing, from improper curing or too high a temperature during fermentation, or from bad growing conditions.
4. **Dark** pickles indicate iron in the water or cooking utensil, ground spices, cooking too long with spices, or too little nitrogen in the cucumbers.
5. **Faded, dull** pickles result from poor growing conditions or too mature cucumbers.
6. **White** sediment in the bottom of jars is not harmful. It could come from not using pure granulated salt, or could be the result of fermentation.

7. **Spoiled** pickles mean you did not process them properly, you used old ingredients, non-standard jars or old lids, or that the pickling solution was not boiling hot, or you filled too many jars before sealing them. In other words, you did not follow directions!

DILLY BEANS

Kids love these beans, and sophisticated adults like them, too — in a martini. Dilly beans can also be a surprise addition to appetizer trays. The recipe makes 6 or 7 (1-pint) jars.

Ingredients	Equipment
4 pounds same-size green beans	*6 or 7 (1-pint) jars with 2-piece self-sealing lids*
3 cups vinegar	*Colander*
3 cups water	*Knife*
1/3 cup pure granulated salt	*Measuring cups*
3/4 to 1 cup dill seed	*Large saucepan*
18 to 21 whole black peppercorns	*Tablespoon*
	Boiling water bath

1. Wash and rinse the jars; keep them hot. Prepare the lids.
2. Wash the beans well; drain. Cut off the ends and trim the beans, if necessary, so they will stand upright in the jars. (If the beans are not the right length to fit in the jar, just trim the ends and cut them into 1- or 2-inch lengths.)
3. Pack the beans into the hot jars. Put 2 tablespoons dill seed and 3 peppercorns into jars.
4. Combine all the remaining ingredients; heat to boiling.
5. Pour the boiling brine into the jars to within ½ inch of each top.
6. Wipe the tops and threads of the jars with a damp cloth.
7. Put on the lids and screw bands.
8. Process in a boiling water bath for 10 minutes. Follow the Basic Steps for Pickles 9 through 16.

PICKLED TINY ONIONS

If you make up your own gift baskets, include a jar of these zesty appetizers. The recipe makes 6 (1-pint) jars.

Ingredients

4 quarts tiny white onions
Boiling water
1 cup pure granulated salt
2 quarts white vinegar
2 cups sugar
3 tablespoons mustard seed
3 tablespoons whole black peppercorns
3 tablespoons grated fresh or prepared horseradish
6 small red pepper pods
3 bay leaves, broken in half

Equipment

Large saucepan
Measuring cups
Measuring spoons
Knife
6 (1-pint) jars with 2-piece self-sealing lids
Colander or sieve
Slotted spoon
Boiling water bath

1. Wash the onions well; drain. Put the onions in the sink or a large pan and pour boiling water over to cover. Let them stand 2 minutes, then drain.
2. Pour cold water over onions to cover. Let them stand just until cool. Drain and peel them.
3. Put the onions in a saucepan or a large mixing bowl, sprinkle with salt and pour cold water over them to cover. Set the onions aside and let them stand overnight.
4. Wash and rinse the jars; keep them hot. Prepare the lids.
5. Drain the onions, then rinse them well with cold water. Set them aside in a colander while preparing the syrup.
6. In a large kettle, combine the vinegar, sugar, mustard seed, peppercorns and horseradish and heat to boiling. Boil 2 minutes.
7. Add the onions and heat to boiling.
8. Spoon the onions into hot jars, packing them gently.
9. Slip a red pepper pod and ½ bay leaf into each jar. Pour in the boiling hot syrup to within ½ inch of each top.
10. Wipe the tops and threads of the jars with a damp cloth.
11. Put on the lids and screw bands.
12. Process in a boiling water bath for 5 minutes. Follow Basic Steps for Pickles 9 through 16.

CORN RELISH

Cabbage, red and green peppers add color and texture to the golden corn. The recipe makes about 6 (1-pint) jars.

Ingredients

- 18 *medium to large ears just-ripe sweet corn*
- 1 *quart chopped cabbage*
- 1 *cup chopped sweet red peppers*
- 1 *cup chopped green pepper*
- 1 *cup chopped onion*
- 1 *to 2 cups sugar*
- 1 *quart vinegar*
- 1 *cup water*
- 1 *tablespoon celery seed*
- 1 *tablespoon mustard seed*
- 1 *tablespoon salt*
- 1 *to 2 tablespoons dry mustard*
- 2 *teaspoons turmeric (optional)*

Equipment

- *6 (1-pint) jars with 2-piece self-sealing lids*
- *Large kettle*
- *Knife*
- *Measuring cups*
- *Measuring spoons*
- *Ladle or large spoon*
- *Wide-mouth funnel*
- *Boiling water bath*

1. Wash and rinse the jars; keep them hot. Prepare the lids as the manufacturer directs.
2. Husk the corn and remove the silk. Cook the ears in boiling water for 5 minutes. Cut the kernels from the cob and measure them. You should have 2 quarts of kernels.
3. Combine the corn and all remaining ingredients in a large preserving kettle.
4. Heat the corn to boiling over high heat, then reduce the heat and simmer 20 minutes, stirring frequently.
5. Ladle, while still boiling, into the hot jars to within ½ inch of each top.
6. Wipe the tops and threads of the jars with a damp cloth.
7. Put on the lids and screw bands as the manufacturer directs.
8. Process in a boiling water bath for 15 minutes. Follow the Basic Steps for Pickles 9 through 16.

PETER PIPER'S PICKLED PUMPKIN

While the kids cut pumpkins for Jack O'Lanterns, you can cut one for these spicy, orange-flavored pickles. The recipe makes 8 to 9 (1-pint) jars.

Ingredients

1 (5-to 6-pound) pumpkin
4 to 5 cups sugar (or 2 to 3 cups sugar, 2 cups honey)
1 quart white or cider vinegar
3 cups water
2 sticks cinnamon
2 (½-inch) chunks fresh ginger root or ¼ cup chopped crystallized ginger
1 tablespoon whole allspice
1 can (6 ounces) frozen orange juice concentrate, thawed

Equipment

8 to 9 (1-pint) jars with 2-piece self-sealing lids
Knife
Measuring cups
Large preserving kettle
Cheesecloth spice bag (optional)
Wooden spoon
Ladle or large spoon
Wide-mouth funnel
Boiling water bath

1. Wash and rinse the jars; keep them hot. Prepare the lids as the manufacturer directs.
2. Wash the pumpkin, cut it in 1-inch chunks and pare. Save the seeds to toast.* You should have about 4 quarts of chunks.
3. In the preserving kettle combine the sugar, vinegar, water and spices. (Tie the spices in cheesecloth, if desired.)
4. Heat over high heat until boiling, stirring constantly.
5. Continue to heat and stir until the sugar dissolves.
6. Stir in the pumpkin chunks and orange juice concentrate and heat to boiling.
7. Reduce the heat to simmer and cook, stirring occasionally, until the pumpkin is just barely tender, about 30 minutes.
8. Ladle into hot jars to within ½ inch of the top, spooning in the hot liquid from kettle, if necessary.
9. Wipe the tops and threads of the jars with a damp cloth.

10. Put the lids and screw bands in place as the manufacturer directs.
11. Process in boiling water bath for 5 minutes. Follow Basic Steps for Pickles 9 through 16.

***TOASTED PUMPKIN SEEDS:** Separate the seeds from strings, spread them on baking sheets, sprinkle with salt and toast in a 375°F oven until golden brown.

SPICED PICKLED PEACHES

If you should have any syrup left over, seal it in clean hot jars to be processed along with the peaches. It can be used for pickling at another time, or as a ham glaze. The recipe makes about 4 (1-quart) or 8 (1-pint) jars.

Ingredients

6 cups sugar
6 cups light corn syrup
1 quart water
1 quart cider vinegar
8 sticks cinnamon
1 tablespoon whole cloves
8 pounds small or medium peaches

Equipment

4 (1-quart) or 8 (1-pint) jars with 2-piece self-sealing lids
Knife
Measuring cups
Measuring spoons
Cheesecloth spice bag
Large preserving kettle
Wide-mouth funnel
Boiling water bath

1. Wash and rinse the jars; keep them hot. Prepare the lids as the manufacturer directs.
2. In large preserving kettle, combine the sugar, syrup, water and vinegar. Tie the spices in a cheesecloth bag and add it to the kettle.
3. Heat over high heat to boil and boil 15 minutes.
4. Meanwhile, wash the peaches. Dip them into boiling water, then into cold water for easier peeling. Put the peeled peaches in cold water to cover and add 2 table-

(Continued On Next Page)

spoons salt and vinegar (or an ascorbic acid mixture) to prevent darkening. Halve and pit the peaches, if desired, or leave them whole.

5. Lift the peaches from the vinegar-water and then gently put a single layer of peaches into the boiling syrup. Simmer 5 minutes or until the peaches are tender. Repeat until all peaches are cooked.
6. Gently pack the peaches into hot jars.
7. Fill the jars with boiling syrup to within ½ inch of each top.
8. Wipe the tops and threads of the jars with a damp cloth.
9. Put on the lids and screw bands as the manufacturer directs.
10. Process in a boiling water bath for 10 minutes. Follow Basic Steps for Pickles 9 through 16.

PICKLED PEARS. Substitute Seckel pears for the peaches, removing the skins and blossom end, but leaving the stems in place. Add 2 tablespoons of whole allspice to the spice bag.

PICKLED WATERMELON RIND. Remove all the green and pink portions of the watermelon rind and cut it in 1-inch chunks. Weigh 8 pounds chunked rind and use it in place of the peaches. Add 1 sliced lemon to the syrup along with the rind and simmer until the rind is clear and transparent.

PICKLED APRICOTS OR CRABAPPLES. Substitute apricots or crabapples for peaches. Wash, but do not peel or cut the fruit.

UNCOMMONLY GOOD CANTALOUPE PICKLES

These are extraordinarily good pickles to bring out for parties or company dinners. The recipe makes 4 (½-pint) jars.

Ingredients

1 medium not-quite-ripe cantaloupe
1 quart vinegar
2 cups water
2 sticks cinnamon
1 tablespoon whole cloves
1 teaspoon ground mace
4 cups brown sugar

Equipment

Knife
Large saucepan
Cheesecloth spice bag
Measuring cups
Measuring spoons
Wooden spoon
Large mixing bowl
4 (½-pint) jars with 2-piece self-sealing lids
Ladle
Wide-mouth funnel
Boiling water bath

1. Peel and seed the cantaloupe; cut it into 1-inch chunks and put them in large mixing bowl.
2. In saucepan combine the vinegar and water. Tie the whole spices in a cheesecloth bag and add it to the saucepan along with the mace. Heat to boiling.
3. Pour the boiling spiced vinegar over the cantaloupe in the mixing bowl. Set the bowl aside and let it stand overnight.
4. The next day, drain vinegar into a saucepan and heat to boiling. Add the cantaloupe and sugar; heat to boiling, then reduce the heat and simmer about 1 hour, or until transparent.
5. Meanwhile, wash and rinse the jars; keep them hot. Prepare the lids as the manufacturer directs.
6. Pack the hot cantaloupe into the hot jars; keep them hot in a low oven or pan of hot water.
7. Boil the vinegar/sugar mixture about 5 minutes or until syrupy.
8. Pour the syrup over the cantaloupe to within ½ inch of each top.
9. Wipe the tops and threads of the jars with a damp cloth.
10. Put on the lids and screw bands.
11. Process in a boiling water bath for 10 minutes.

EAST INDIA RELISH

Stir a few tablespoons of this relish into mayonnaise when making a chicken salad, or try it with pot roast. The recipe makes 6 (1-pint) jars.

Ingredients

- 1½ *quarts chopped onion (about 6 to 8 medium onions)*
- 1 *quart firmly-packed shredded carrots (about 1 pound)*
- 1 *quart peeled, diced green tomatoes (about 8 medium)*
- 1 *quart chopped zucchini (about 2 pounds)*
- 2 *cups light corn syrup*
- 2 *cups white vinegar*
- 2 *tablespoons salt*
- 1 *tablespoon ground coriander*
- 2 *teaspoons ground ginger*
- 1 *teaspoon crushed red pepper*
- ½ *teaspoon ground cumin*

Equipment

- *6 (1-pint) jars*
- *Knife*
- *Measuring cups*
- *Measuring spoons*
- *Large preserving kettle*
- *Wooden spoon*
- *Ladle*
- *Wide-mouth funnel*
- *Boiling water bath*

1. Wash and rinse the jars; keep them hot. Prepare the lids as the manufacturer directs.
2. In a preserving kettle, combine the ingredients.
3. Heat over medium-high heat to boiling.
4. Reduce the heat and simmer 5 minutes.
5. Spoon the relish into hot jars until each jar is almost full.
6. Spoon hot liquid from the kettle to within ¼ inch of top of the jars.
7. Wipe the tops and threads of the jars with a damp cloth.
8. Put on the lids and screw bands as the manufacturer directs.
9. Process in a boiling water bath for 10 minutes. Follow Basic Steps for Pickles 9 through 16.

PLUM AND PEACH CHUTNEY

You have many possible variations for this heavily-spiced chutney. Use peaches and plums, or substitute nectarines or apricots. The recipe makes 5 to 6 (½-pint) jars.

Ingredients

- *1½ pounds peaches*
- *1½ pounds plums*
- *1 pound brown sugar*
- *2 cups white vinegar*
- *¼ cup shredded fresh ginger root or chopped crystallized ginger*
- *1 clove garlic, minced or ½ teaspoon garlic powder*
- *2 teaspoons cinnamon*
- *2 teaspoons ground cloves*
- *1 teaspoon salt*
- *¼ teaspoon pepper*
- *1 cup chopped onion*
- *1 cup chopped green pepper*

Equipment

- *6 (½-pint) jars with 2-piece self-sealing lids*
- *Measuring cups*
- *Measuring spoons*
- *Large saucepan*
- *Pot or preserving kettle*
- *Knife*
- *Wooden spoon*
- *Ladle*
- *Wide-mouth funnel*
- *Boiling water bath*

1. Wash and rinse the jars; keep them hot. Prepare the lids as the manufacturer directs.
2. Wash the plums and peaches. Halve, remove the pits and chop them. You should have 6 cups of chopped fruit to set aside.
3. In a large pot or kettle combine the sugar, vinegar, ginger, garlic and spices. Heat to boiling, stirring constantly.
4. Add the onions, peppers, peaches and plums.
5. Heat to boiling over high heat, then reduce the heat slightly and boil gently over medium heat 1 to 1¼ hours or until very thick, stirring occasionally.
6. Ladle into hot jars to within ½ inch of each top.
7. Wipe the tops and threads of the jars.
8. Put on the lids and screw bands as the manufacturer directs.
9. Process in a boiling water bath 5 minutes. Follow Basic Steps for Pickles 9 through 16.

APPLE-GREEN TOMATO CHUTNEY

Want an instant appetizer? Put cream cheese on a pretty plate and spoon this chutney over it. Pass the crackers. Delicious. The recipe makes about 10 (1-pint) jars.

Ingredients

1 dozen medium tart apples
6 medium green tomatoes
1 cup chopped onions
1 cup chopped green and/or sweet red peppers
1 pound raisins or currants
1 quart cider vinegar
3 cups brown sugar
3 tablespoons mustard seed
2 tablespoons ground ginger
2 teaspoons salt
2 teaspoons ground allspice

Equipment

Knife
Measuring cups
Measuring spoons
Large preserving kettle
Wooden spoon
10 (1-pint) jars with 2-piece self-sealing lids
Ladle or large spoon
Wide-mouth funnel
Boiling water bath

1. Wash the fruits and vegetables. Core the apples, stem the tomatoes, then chop them coarsely. In large preserving kettle, combine the apples and tomatoes with all other ingredients.
2. Heat to boiling then reduce heat and simmer about 1 to 1¼ hours, stirring occasionally.
3. Meanwhile, wash and rinse the jars; keep them hot. Prepare the lids as the manufacturer directs.
4. Ladle the boiling hot chutney into the hot jars to within ½ inch of the top.
5. Wipe the tops and threads of the jars with a damp cloth.
6. Put on lids and screw bands as the manufacturer directs.
7. Process in a boiling water bath for 5 minutes. Follow Basic Steps for Pickles 9 through 16.

ORCHARD AND VINEYARD CATSUP

Use all plums or all grapes, if you wish, or try almost any other fruit for this rich catsup — peaches, pears, nectarines, apricots, apples, seeded red or black grapes. The recipe makes 4 (½-pint) jars.

Ingredients

3 pounds ripe plums
3 pounds seedless grapes
3 cups water
Sugar (or 2/3 sugar, 1/3 corn syrup or honey)
3 cups vinegar
1 to 2 tablespoons pumpkin pie spice OR 1 teaspoon each ground cinnamon, cloves, nutmeg and allspice

Equipment

Large saucepan or kettle
Sieve or food mill
Preserving kettle
Measuring cups
Measuring spoons
4 (½-pint) jars with 2-piece self-sealing lids
Ladle
Wide-mouth funnel
Boiling water bath

1. Wash the plums and grapes; stem the grapes.
2. Put the fruit in a large saucepan with water and simmer about 30 minutes or until very soft.
3. Press the fruit through a sieve or food mill; discard the skins and pits.
4. Measure the puree into a large preserving kettle. Then measure ½ as much sugar as puree into the kettle.
5. Stir in the vinegar and spices.
6. Heat to boiling, then reduce the heat and simmer about 1 hour and 15 minutes or until very thick. Skim the foam, if necessary.
7. Wash and rinse the jars; keep them hot. Prepare the lids as the manufacturer directs.
8. Ladle the catsup into hot jars to within ½ inch of the top.
9. Wipe the tops and threads of the jars with a damp cloth.
10. Put on the lids and screw bands as the manufacturer directs.
11. Process in a boiling water bath for 10 minutes. Follow Basic Steps for Pickling 9 through 16.

KATHY'S CATSUP

Tangy and fragrant, this catsup is easy to make if you use a blender to simplify the preparation. The recipe makes about 5 pints.

Ingredients

1 peck (8 quarts) ripe tomatoes
3 onions
2 dried red peppers, chopped
OR
1 teaspoon dried peppers
1½ bay leaves
1 tablespoon whole allspice
1 clove garlic, peeled
1 stick cinnamon
2 cups vinegar
½ cup sugar
1 tablespoon celery salt

Equipment

Knife
Blender
Measuring cups
Strainer
Large mixing bowl
Large preserving kettle
Cheesecloth for spice bag
Wooden spoon
Slotted spoon
5 (1-pint) jars with 2- piece self-sealing lids
Ladle
Wide-mouth funnel
Boiling water bath

1. Wash, stem and quarter the tomatoes. Peel and quarter the onions.
2. Fill the blender container almost to the top with the tomatoes and blend until smooth.
3. Set the strainer over a large bowl and pour the blended tomatoes through it. Repeat the blending and straining for the remaining tomatoes, including the onion quarters added to the last batch of tomatoes.
4. Measure 4 quarts of puree into a large preserving kettle.
5. Tie the spices in a cheesecloth bag and add it to the kettle along with all the remaining ingredients.
6. Heat to boiling over high heat, stirring constantly. Then reduce the heat slightly and boil until thick, about 1 hour, stirring frequently. Reduce the heat and stir more often during end of the cooking time to prevent sticking.
7. Skim the foam with a slotted spoon, if necessary.
8. Meanwhile, wash and rinse the jars; keep them hot. Prepare the lids as the manufacturer directs.

9. Wipe the tops and threads of the jars with a damp cloth.
10. Ladle the hot catsup into the hot jars to within ¼ inch of each top.
11. Put on the lids and screw bands as the manufacturer directs.
12. Process in a boiling water bath for 10 minutes. Follow the Basic Steps for Pickling 9 through 16.

SAUERKRAUT

Creamy-white and tangy, kraut is fermented cabbage. Pick a cool, out-of-the-way spot to put the crock while the kraut ferments. The recipe makes 16 to 18 quarts.

Ingredients

40 pounds cabbage
1 pound (1½ cups) pure granulated salt

Equipment

Sharp knife or kraut shredder
Large mixing bowl or other large, clean nonmetallic container
Scale
Measuring spoons
Wooden spoon
Large (at least 5 gallons) clean crock or other large nonmetallic container
Heavy-duty, large plastic food bag
Large kettle
16 to 18 (1 quart) jars with 2-piece self-sealing lids
Ladle
Wide-mouth funnel
Boiling water bath

1. Choose firm, mature heads of cabbage and remove the outer leaves. Cut out any bad portions.
2. Wash the cabbage well and drain.
3. Cut the heads into halves or quarters and cut out the cores.

(Continued On Next Page)

4. Shred with a kraut shredder (or sharp knife) into shreds no thicker than a dime.
5. Weigh 5 pounds of shredded cabbage and put it into a large mixing bowl or other nonmetallic container.
6. Sprinkle the cabbage with 3 tablespoons salt and mix well; let it stand several minutes until slightly wilted.
7. Firmly pack the salted cabbage in even layers in a large clean crock or other nonmetallic container (such as a jar, or a brand new plastic wastebasket or garbage can).
8. Repeat weighing 5 pounds of cabbage and mixing it with 3 tablespoons salt, packing into crock, until all the cabbage is used or until the cabbage is packed to within 3 or 4 inches of the top of the crock.
9. Press down firmly on the shredded cabbage in the crock until liquid comes to the surface.
10. Put a heavy-duty food bag on top of the kraut in the crock and pour enough water into the bag to completely cover the surface of the kraut in the crock. Add enough more water to provide sufficient weight to hold the cabbage underneath the brine. Tie the bag with a twist tie, rubber band or string.
11. Set the crock in a cool place (68°F to 72°F) for 5 to 6 weeks or until the mixture has stopped bubbling. The kraut should be creamy-white, mildy tart and tangy.
12. When the kraut has stopped fermenting transfer it, several quarts at a time, to a large kettle and heat just to simmering, but do not boil.
13. Pack the hot sauerkraut into hot quart jars to within ½ inch of each top.
14. Pour in the hot sauerkraut juice (from the kettle) to within ½ inch of each top.
15. Wipe the tops and threads of the jars with a damp cloth.
16. Put on the lids and screw bands as the manufacturer directs.
17. Process in a boiling water bath, 15 minutes for pints, 20 minutes for quarts. Start to count the processing time as soon as the jars go into the boiling water. Follow Basic Steps for Pickles 9 through 16.

Steps 4 and 5: Shred cabbage into shreds no thicker than a dime before weighing out 5 pounds of it.

Step 6: Sprinkle cabbage with 3 tablespoons salt, mix well and let it stand several minutes.

Step 7: Firmly pack slightly wilted cabbage into clean crock, plastic bucket or other nonmetalic container.

Step 10: Put large, heavy-duty plastic bag on cabbage, and fill with enough water to cover the cabbage's surface.

Step 12: When kraut stops fermenting ladle it into a large kettle and heat to simmering.

Step 13: Pack hot kraut into hot jars to within 1/2 inch of each top.

Step 14: Pour in hot kraut juice to within 1/2 inch of top.

Step 15: Wipe the tops and threads of jars with a damp cloth.

Step 16: Put on lids as the manufacturer directs.

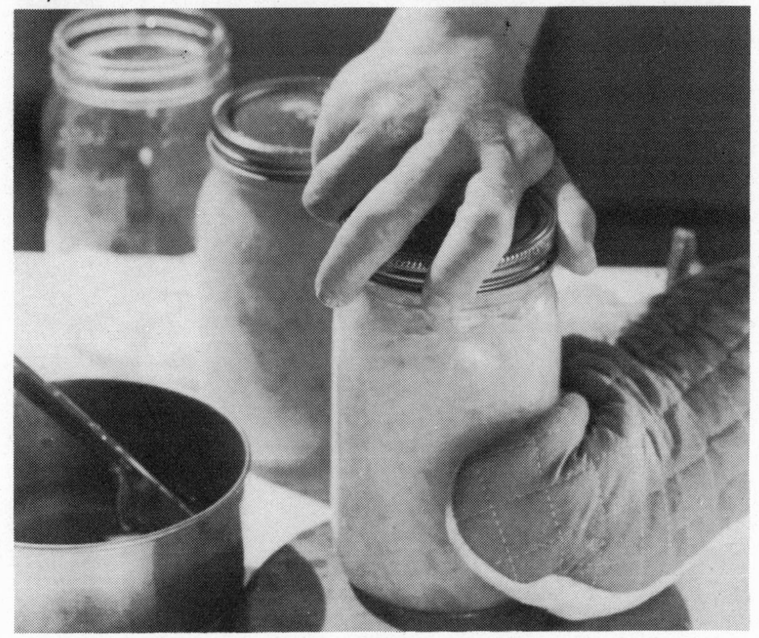

Step 16: Put on the screw bands as the manufacturer directs.

Step 17: Process in a boiling water bath. Follow Basic Steps for Pickles 9 through 16.

SAUERKRAUT ROUBAL

Mrs. Anton Roubal of Bruno, Nebraska, makes a tasty dish from her kraut by sauteeing a medium chopped onion in fat, then adding the kraut and some meat juice along with a tablespoon of brown sugar, ½ teaspoon caraway seed and salt to taste. After it simmers for an hour she thickens it with 2 tablespoons flour mixed with ¼ cup cold water.

Blue Ribbon Recipes

Crunchy dills, zesty relishes, juicy pickled vegetables — take your choice, they are all award-winning recipes from State and county fairs across the land.

The great tradition of American fairs started at the beginning of the 19th century. The first State-run fair was the New York State Fair held at Syracuse in 1841. State-organized fairs replaced fairs run by agricultural societies, spreading westward. New Jersey held a fair shortly after New York. Michigan followed next with a fair in 1849. The first State fairs in Pennsylvania, Ohio and Wisconsin were held in 1851, followed by Indiana in 1852, Illinois in 1853, and Iowa in 1854.

BREAD AND BUTTER PICKLES U.S.A.

These fresh-packed, sliced pickles must be a nation-wide favorite. We received more prize-winning recipes for Bread and Butter Pickles (sometimes called Crispie Lunch Pickles, or Crunchy Cucumber Pickles) than any other recipe. The following people won prizes for this recipe, or one very similar. Eloie Wilson won a blue ribbon at the Colorado State Fair for her pickles; Glenn A. Welch took first place at the Montana Winter Fair for his recipe. In the South, Dale Lollard was the State winner in the 4-H Alabama Food Preservation Project; and Vicki Childs took a blue ribbon at the Mitchell County Fair in Georgia. Moving up the Atlantic Coast, we find Vicky Huffman from Maryland using a similar recipe to help win a trip to the National 4-H Congress; a 4-H'er from New Jersey, Nancy Gottlick won at the Somerset County Fair; and Mrs. Ronald MacDougal sent a winning recipe from the Northern Maine Fair. The recipe makes about 9 (1-pint) jars.

(Continued On Next Page)

6 pounds medium cucumbers (about 1 gallon slices)
6 medium white onions
4 green peppers, chopped
3 cloves garlic, peeled (optional)
1/3 cup pure granulated salt
1 tray ice cubes
5 cups sugar
3 to 4 cups vinegar
2 tablespoons mustard seed
1½ teaspoon turmeric
1½ teaspoons celery seed

Slice the cucumbers and onions very thin. Put them in a large bowl or plastic dishpan; add the green peppers and garlic cloves. Sprinkle with salt and mix well. Cover the vegetables with cracked ice or ice cubes and let them stand for 3 hours. Remove any ice; drain well. In a large preserving kettle, combine the sugar, vinegar and spices; heat to boiling. Add the drained vegetables and heat to boiling. Ladle them into clean, hot pint jars to within ½ inch of each jar's top. Wipe off the tops and threads of the jars with a damp cloth. Put on the prepared lids and seal as the manufacturer directs. Process in a boiling water bath 10 minutes. Let the pickles stand four weeks before using.

VARIATIONS: Mrs. Gene Kirkman sent a Bread and Butter Pickles recipe that uses 2 green mango peppers, sliced, in place of the garlic. She also adds ½ teaspoon of whole cloves and suggests you add 2 hot peppers for really hot pickles. Her version won five blue ribbons at Ohio State Fairs and a Grand Award at the Franklin County Fair!

Mary Besich adds a stick of cinnamon to her variation, or a can of pimientoes, chopped. She also suggests 2 tablespoons dill seed and sometimes uses sliced zucchini instead of cucumbers. The result is a tasty Dilled Zucchini.

Mrs. Lucius Nix won with her recipe at the North East Georgia Fair. She uses very large cucumbers, including those that have started to turn yellow. She peels and seeds the cukes and cuts them into long strips, soaks them for a day in lime water, then drains them. Next, she pours a boiling vinegar-sugar-spice mixture over the cucumber sticks and lets them stand overnight. The following day she simmers the sticks in the syrup until transparent, then packs them into jars, seals and processes as above.

CAULIFLOWER PICKLES MACDOUGAL

Here is another blue ribbon winner from the Northern Main Fair, Presque Isle, prepared by Mrs. Ronald MacDougal. These crisp, sweet-sour caulifloweretes are a must in any pickle dish. The recipe makes 4 (1-pint) jars.

3 quarts caulifloweretes (about 1½ to 2 large heads)
1 cup vinegar
1 cup water
3 cups sugar
3 cups vinegar
3 tablespoons mixed pickling spice

In a large saucepan or preserving kettle, cook the caulifloweretes in the 1 cup water and 1 cup vinegar until just crisp-tender, about 5 minutes. While the cauliflower cooks, combine the sugar and vinegar in a large saucepan. Tie the spices in a cheesecloth bag and add them to the sugar and vinegar. Heat to boiling, then boil 5 minutes. Remove spice bag. Drain the caulifloweretes well, then pack them tightly into clean, hot pint jars. Pour the hot syrup over the cauliflower to within ½ inch of each jar's top. Wipe off the tops and threads of the jars with a damp cloth. Put on prepared lids and seal as the manufacturer directs. Process in boiling water bath for 5 minutes.

CRISP ZUCCHINI PICKLES STONEHOUSE

Dorothy Stonehouse from the Spragueville Extension at Presque Isle, Maine, pickles zucchini along with several other garden favorites. You could use whole green cherry tomatoes in place of the green beans if you wish. The recipe makes about 8 (1-pint) jars.

5 quarts cubed zucchini
1 quart sliced onion
1 large head cauliflower, separated into flowerets
2 green peppers, seeded and chopped
1 quart trimmed and chopped green beans or peeled and sliced carrots
3 cloves garlic, halved
½ cup salt
6 cups sugar
5½ cups vinegar
½ cup water
2 tablespoons mustard seed
2 teaspoons celery seed
2 teaspoons turmeric

In a large mixing bowl or crock, combine all the vegetables, including the garlic. Sprinkle them with the salt; stir and let the vegetables stand 3 hours. Drain well.

Combine the sugar, vinegar, water and spices in a large preserving kettle and heat to boiling. Stir in the vegetables and heat again to boiling. Pack the vegetables into clean, hot pint jars to within ½ inch of each jar's top. Pour in the hot liquid to within ½ inch of each top. Wipe off the tops and threads of the jars with a damp cloth. Put on prepared lids and seal as the manufacturer directs. Process in a boiling water bath for 15 minutes.

DILL PICKLES

Bonnie L. Michaels and Dorothy C. Davis, both of Illinois, won blue ribbons for their fresh packed dills. These juicy pickles "pickle" in the jar — what could be easier! The recipe makes about 6 to 7 quart jars.

8 pounds medium (3 inch) cucumbers (5 to 7 per quart jar)
1 quart cider vinegar
2 quarts water
¾ cup pure granulated salt
About 12 heads or sprigs fresh dill
12 small cloves garlic, peeled

Wash the cucumbers well, being sure to remove the blossom ends. Heat the vinegar, water and salt to boiling. Pack the cucumbers gently but firmly into clean, hot quart jars, putting 2 heads of dill and 2 garlic cloves in each jar. Pour in the boiling liquid to fill the jars and let them stand for 15 minutes. Pour the liquid from the jars back into the pan and heat again to boiling, then pour it into the jars to within ½ inch of each top. Wipe off the tops and threads of jars with a damp cloth. Put on prepared lids and seal as the manufacturer directs. Process in a boiling water bath for 20 minutes.

MOM'S TOMATO PICKLES

Those green tomatoes left in the garden at the first frost can be made into snappy-sweet pickles, as Mrs. Ronald MacDougal's winning recipe from the Northern Maine Fair proves. The recipe makes about 5 (1-pint) jars.

About 2 dozen medium green tomatoes
8 to 10 large onions, sliced
1 cup pure granulated salt
3 quarts vinegar
4 cups brown sugar
3 tablespoons mixed pickling spice (tied in cheesecloth bag)

Stem and slice the tomatoes. Layer the tomato and onion slices in a large bowl or crock with the salt. Let them stand overnight. Drain well. In a large preserving kettle, heat the vinegar, brown sugar and spices to boiling. Add the drained vegetables and simmer 15 minutes or until the tomatoes are tender. Pack into clean, hot jars to within ½ inch of each top. Run a slim, nonmetal tool down side of the jars to release air bubbles. Add additional liquid, if necessary. Wipe off the tops and threads of the jars with a damp cloth. Put on prepared lids and seal as the manufacturer directs. Process in a boiling water bath for 5 minutes.

MIXED VEGETABLE PICKLES HUGHES

There is a lot of cutting and chopping to Connie Hughes' recipe, but one taste of these crisp-tender, tangy vegetables will make the effort seem worthwhile. Mrs. Hughes, from St. John, Washington, invented this recipe and it brought her one of the many blue ribbons she won at the Palouse Empire Fair. Her directions call for measurement of the cut vegetables by weight, so bring out your kitchen scale. The recipe makes about 12 (1-pint) jars.

- *1 pound sweet red pepper strips (cut 1 x ½ inch), about 3 large peppers*
- *1 pound green pepper strips (cut 1 x ½ inch), about 4 medium peppers*
- *2 pounds cauliflowerets, about 1 to 1½ inches*
- *¾ pound carrots sliced ¼ inch thick, about 9 medium carrots*
- *1 pound onion slices ½ inch thick, cut in 1 inch pieces or small white onions, peeled*
- *7 cups water*
- *7 cups white vinegar*
- *½ cup pure granulated salt*

Mix all the vegetables thoroughly in a very large container (a dishpan or clean sink works well). Spoon or ladle them into clean, hot pint jars to within ½ inch of each jar's top, pressing the vegetables firmly into each jar. Heat the water, vinegar and salt to boiling and pour over the vegetables in the jars. Run a slim, nonmetal tool down the side of each jar to release air bubbles, then pour in additional brine to within ½ inch of each jar's top. Wipe off the tops and threads of the jars with a damp cloth. Put on prepared lids and seal as the manufacturer directs. Process in a boiling water bath for 15 minutes. Let the pickles stand at least 4 weeks before eating, to complete the pickling.

NOTE: Should you need more brine, Mrs. Hughes has worked out the proportions for you.

3½ cups water, 3½ cups vinegar plus ¼ cup salt
1¾ cup water, 1¾ cup vinegar plus 2 tablespoons salt
⅞ cup water, ⅞ cup vinegar plus 1 tablespoon salt

SWEET PICKLE STICKS

Glenn A. Welch of Bozeman, Montana, took the pickle sweepstakes at the Montana Winter Fair with his recipe. Mrs. George Sanders, who lives in Illinois, halfway across the country from Mr. Welch, has a prize-winning variation. She uses small whole cukes (stuck with a fork so syrup can soak in) or chunks of larger cucumbers. If you own large crocks and a really big kettle you can double this recipe. The recipe makes about 4 quart jars.

1 gallon medium (3 inch) cucumbers
2 cups pure granulated salt
1 gallon boiling water
1 gallon boiling water
1 gallon boiling water
1 tablespoon alum
16 cups sugar (6 pounds)
2 quarts vinegar
½ cup mixed pickling spice

Wash the cucumbers well and put them in a large crock (at least 3 gallons). Pour the salt over them; then pour the first gallon of boiling water over the cukes. Cover and set them aside in a cool place for 1 week. Drain off the brine. Pour on the second gallon of boiling water and let them stand overnight. Drain. Stir the alum into the last gallon of boiling water and pour it over the cukes. Let them stand overnight. Drain. Heat the sugar, vinegar and spices together and pour them over the pickles. Let them stand overnight. Drain and reserve the syrup and heat it to boiling.

Pack the pickles in clean, hot quart jars to within ½ inch of each top. Pour in the boiling syrup to within ½ inch of each top. If necessary, run a slim nonmetal tool down the sides of the jars to release any air bubbles; add additional syrup, if necessary. Wipe off the tops and threads of the jars with a damp cloth. Put on prepared lids and seal as the manufacturer directs. Process in a boiling water bath for 5 minutes. Chill before serving, says Mr. Welch.

PICKLED BEETS

Use small, young beets for these sweet-sour vegetables, or halve or quarter larger beets. Vera Roody of Corning, New York, won a blue ribbon at the New York State Fair for this recipe. And, Eloie Wilson won a blue ribbon at the Colorado State Fair using this recipe. Whether you live in the East or West, you will want to serve these beets when you make a Scandanavian-style dinner. The recipe makes about 3 (1-pint) jars.

2 dozen small beets
2 cups sugar
2 cups vinegar (or half vinegar and half water for less tart beets)
1 teaspoon salt

Cut the tops off of the beets, leaving 1 inch of stem. Wash them well. Put the beets in a large saucepan or pot and add enough water to cover. Add 1 tablespoon salt for each quart water. Cover and cook until tender. While the beets are cooking, make a syrup by heating the sugar, vinegar and salt to boiling. When the beets are tender, drain and cool in cold water. Cut off the tops and roots; slip off the skins. Pack whole small beets or quartered larger beets into clean, hot pint jars. Pour in boiling syrup to within ¼ inch of each top. (If you need more syrup just combine sugar and vinegar in equal amounts.) Wipe off the tops and threads of the jars with a damp cloth. Put on prepared lids and seal as the manufacturer directs. Process in a boiling water bath for 10 minutes.

PICKLE PEPPERS KITTLE

Loyd Edward Kittle from Ringgold, Georgia, writes that "Yes, I was indeed proud to win the Blue Ribbon, as first place winner, in the Salem Valley Fair." Here is his prize-winning condiment. He suggests that, for added flavor, you can also use small onions, small green tomatoes and/or mixed pickling

spices. His recipe gives the proportions for 1 quart; you can increase the recipe to match the productivity of your pepper plants. Warning: These are HOT! Recipe makes 1 quart jar.

4 medium sweet red peppers
4 medium green peppers
10 banana peppers
1 cup vinegar
½ cup water
1 teaspoon salt
1 teaspoon sugar

Stem and seed all the peppers. Cut the larger peppers into thirds or large chunks. Cut the banana peppers into halves or thirds. Heat the vinegar, water, salt and sugar in a large saucepan or kettle. Add the peppers and heat only until warm. Lift the peppers from the liquid and pack them into the jars with the red peppers on the outside. Heat the liquid to boiling and pour it over the peppers to within ½ inch of each top. Wipe off the tops and threads with a damp cloth. Put on prepared lids and seal as the manufacturer directs. Process in a boiling water bath for 5 minutes.

CUCUMBER RELISH JOHNSON

Jodi Johnson's winner has a golden touch — carrots. This relish won a blue ribbon at the Red River Valley Fair in North Dakota. You will need a grinder for this recipe. The recipe makes about 8 (1-pint) jars.

4 pounds cucumbers (about 20 medium)
¼ cup pure granulated salt
6 carrots
4 onions
2 green peppers
1 cup chopped pimiento
4 cups sugar
3 cups vinegar
1 teaspoon mustard seed
1 teaspoon celery seed
1 teaspoon turmeric

Grind the cucumbers. Mix them with salt and let them stand for 3 hours. Drain well. Grind the carrots, onions and peppers together. Mix them with the drained cucumbers and all the remaining ingredients in large preserving kettle; heat to boiling. Simmer 20 minutes. Ladle the relish into clean, hot pint jars to within ½ inch of each top. Wipe off the tops and threads of the jars. Put on prepared lids and seal as the manufacturer directs. Process in a boiling water bath for 10 minutes.

SUNSHINE PICKLES BESICH

Mary J. Besich, a prize-winning member of the Cascade County, Montana, Extension Homemaker Club, sent us her recipe for flavorful pickle slices. The recipe makes about 7 (1-pint) jars.

About 4 pounds cucumbers
4 pounds, or about 12 onions
1 cup salt
Cold water
5 cups sugar
1 quart vinegar
1 can (4 ounces) pimiento, chopped
2 tablespoons mustard seed
2 teaspoons turmeric
⅛ teaspoon alum (optional)

Wash, peel and thinly slice the cucumbers. You should have about 6 quarts of slices. Peel and slice the onions. Layer the cucumber and onion slices with the salt in a large bowl. Add enough cold water to cover. Let it stand 2 hours. Drain well.

Put all the remaining ingredients into a large preserving kettle and heat to boiling, stirring to dissolve the sugar. Add the drained cucumbers and onions and heat to boiling; boil 5 minutes. Spoon the hot pickles into clean, hot pint jars to within ½ inch of each jar's top. Pour in the hot liquid from the kettle to within ½ inch of each jar's top. Run a slim nonmetal tool down the sides of the jars to release air bubbles, then pour in additional liquid, if needed. Wipe the tops and threads of the jars with a damp cloth. Put on prepared lids and seal as the manufacturer directs. Process in a boiling water bath for 5 minutes.

SPICED CRAB APPLES HILDRETH

There is no prettier garnish for a plattered roast than these bright preserves. Mrs. Charles Hildreth won an Ohio State Fair blue ribbon with her recipe. The recipe makes 5 to 6 (1-pint) jars.

About 2 quarts evenly sized, sound crab apples
1 quart cider vinegar
4 cups sugar
3 cups water
1 tablespoon whole cloves
1 stick cinnamon
1 teaspoon whole allspice
1 teaspoon ground mace

Wash the apples but do not peel. Tie the spices in a cheesecloth bag. In a preserving kettle, combine all the remaining ingredients and heat to boiling. Cool the syrup to room temperature, then add the apples and heat very slowly just to below simmering, being careful not to burst the skins. Remove from heat and let the apples stand in syrup overnight. Remove the spice bag and pack the apples into clean, hot pint jars to within ½ inch of each jar's top. Pour in the syrup to within ½ inch of each jar's top. Wipe off the tops and threads of the jars with a damp cloth. Put on prepared lids and seal as the manufacturer directs. Process in a simmering water bath for 20 minutes.

SWEET FRANKFURTER RELISH BESICH

A blue-ribbon winner by Mary S. Besich from Black Eagle, Montana, this sweet-sour relish is marvelously easy to make. You will need a food grinder. The recipe makes 2 to 3 (½-pint) jars.

2 medium carrots, peeled
2 medium tomatoes, peeled and quartered
1 medium cucumber, seeded and quartered
2 cups sugar
2 cups vinegar
1 tablespoon mixed pickling spice in cheesecloth bag
1 teaspoon salt
⅛ teaspoon cayenne pepper

Grind the vegetables together; drain. Combine the vegetables with all remaining ingredients in large saucepan and simmer 45 minutes, stirring occasionally. Remove the spice bag. Ladle the relish into clean, hot ½-pint jars to within ½ inch of each top. Wipe off the tops and threads of the jars with a damp cloth. Put on prepared lids and seal as the manufacturer directs. Process in a boiling water bath for 5 minutes.

DUTCH SALAD SPRAGUE

Alice Sprague of Presque Isle, Maine, thickens her champion end-of-the-garden relish with a little flavor in the old-fashioned manner. Turmeric adds flavor and sunshine color. Dutch Salad is popularly served with Maine Number 1 Red Hot Dogs — a unique down east specialty. The recipe makes about 13 (1-pint) jars.

- *1 head (about 2 pounds) cauliflower, separated into small flowerets*
- *1 head (about 2 pounds) cabbage, coarsely chopped*
- *1 bunch celery, coarsely chopped*
- *1 quart coarsely chopped green tomatoes*
- *1 quart coarsely chopped cucumber*
- *1 quart chopped onions*
- *3 sweet red peppers, chopped*
- *1 gallon water*
- *1 cup pure granulated salt*
- *3 cups sugar*
- *1 cup flour*
- *1 cup vinegar*
- *3 pints (6 cups) white or cider vinegar*
- *1 pint water*
- *2 tablespoons celery seed*
- *2 tablespoons mustard seed*
- *1 tablespoon turmeric*

Put all the vegetables in large bowl or container. Combine the 1 gallon water and salt and stir until the salt dissolves. Pour the salt-water mixture over the vegetables and let them stand overnight. The next morning, drain the vegetables well. Stir the sugar and flour together, then mix in the 1 cup vinegar until smooth. Stir in all the remaining ingredients and heat to boiling. Add the drained vegetables, heat to boiling then lower the heat and simmer 20 minutes, stirring frequently. Ladle the relish into clean, hot pint jars to within ½ inch of each jar's top. Wipe off the tops and threads of the jars with a damp cloth. Put on prepared lids and seal as the manufacturer directs. Process in a boiling water bath for 20 minutes.

NOTE: Unless you have a very large kettle, you may need to simmer the vegetables, vinegar and spices in two batches,

cooking about 3 quarts chopped vegetables and about 5 cups vinegar-spice mixture for each batch.

VARIATIONS: Dutch Salad is only one version of the mustard pickle, a specialty of Aroostock County, Maine. Dorothy Stonehouse's recipe, called Thick Mustard Pickles, is another version. For Thick Mustard Pickles you use green tomatoes, small onions and cauliflower.

Another winning mustard pickle recipe comes from Mrs. Ronald MacDougal. Named Mom's Mustard Paste Pickles, it calls for four cups brown sugar instead of three cups white sugar, because Mrs. MacDougal likes a sweeter pickle.

CHOW CHOW SMITH

Ann Smith of Powder Springs, Georgia, got this blue ribbon recipe from her mother-in-law. If you have a lot of green tomatoes and a very large preserving kettle you can double the recipe. The recipe makes about 11 (1-pint) jars.

- 6 *pounds green tomatoes, chopped*
- 3 *cups boiling water*
- 1 *cup salt*
- 5 *large onions, chopped*
- 3 *green peppers, chopped*
- 3 *sweet red peppers, chopped*
- 1 *head cabbage, coarsely chopped*
- 3 *medium cucumbers, chopped*
- 3 *medium carrots, chopped*
- 1 *bunch celery, chopped*
- 1 *quart vinegar*
- 2 *cups sugar*
- ½ *cup dry mustard*
- 1 *tablespoon turmeric*

Put the chopped tomatoes into a large bowl. Stir the salt into the boiling water until it dissolves, then pour it over the tomatoes. Let them stand for 3 hours. Drain and mix the tomatoes with the remaining vegetables, except the celery. Heat the vinegar, sugar, mustard and turmeric to boiling in a large preserving kettle. Add the vegetables (except celery) and simmer for 10 minutes. Stir in the celery. Ladle the relish into clean, hot pint jars to within ½ inch of each jar's top. Wipe off the tops and threads of the jars with a damp cloth. Put on prepared lids and seal as the manufacturer directs. Process in a boiling water bath for 15 minutes.

PEPPER RELISH CROWE

This is an outstanding topper for burgers, hot dogs or sandwiches. Pepper Relish won a prize at the North Georgia State Fair for Frankie Crowe of Dallas, Georgia. The recipe makes about 6 (½-pint) jars.

12 sweet red peppers
12 green peppers
6 medium onions
3 small hot peppers
1 cup water
1 cup salt
2 cups vinegar
1 cup brown or granulated sugar
3 tablespoons salt
1 tablespoon mixed pickling spice

Cut the peppers in half lengthwise and remove the seeds. Arrange them in a bowl or pan and cover with a brine made from the water and the 1 cup salt. Set the bowl aside for several hours or overnight. Drain well and put the peppers through a food grinder with the onions. Put the peppers and onions in a large preserving kettle; add boiling water to cover and let it stand 10 minutes. Drain well. Add boiling water to cover a second time and let it stand 10 minutes. Drain well. Return the peppers and onions to the kettle; add the remaining ingredients and heat to boiling. Reduce the heat and simmer 15 minutes. Pack into clean hot ½-pint jars to within ½ inch of each jar's top. Wipe off the tops and threads of the jars with a damp cloth. Put on prepared lids and seal as the manufacturer directs. Process in a boiling water bath for 15 minutes.

PICKLE RELISH HIGGINS

Cheryl Higgins, a blue-ribbon winner from Plover, Wisconsin, went to the National 4-H Congress Food Preservation. Here is her prize-winning relish. She gives pimientos as an alternate ingredient if red peppers aren't available. The recipe makes about 9 (1-pint) jars.

15 to 20 large (4 to 5 inch) cucumbers (5 to 6 pounds)
4 or 5 onions
1 bunch celery
1 green pepper
1 sweet red pepper
2 green tomatoes
½ cup pure granulated salt
1 quart white vinegar
4 cups sugar
2 tablespoons mustard seed
1 tablespoon celery seed
1½ teaspoons turmeric powder
1½ teaspoons ground cloves

Grind the cucumbers, onions, celery, peppers and tomatoes. Sprinkle with salt and let them stand at least 1 hour. Drain well. Combine all the remaining ingredients in a large preserving kettle. Add the drained vegetables and simmer, stirring constantly, about 15 minutes or until the mixture loses its green color. Ladle the relish into clean, hot pint jars to within ½ inch of each top. Wipe off the tops and threads of the jars with a damp cloth. Put on prepared lids and seal as the manufacturer directs. Process in a boiling water bath for 10 minutes.

TANGY TOMATO RELISH CANTRELL

Sweet, sour, hot and flavorful — all those words describe Betty Cantrell's recipe. It was a winner at Cobb County's North Georgia State Fair and Mrs. Cantrell says the relish is really hot, so use fewer peppers if you see fit. The recipe makes about 4 (1-pint) jars.

6 pounds green tomatoes, chopped
4 large onions, chopped
4 sweet red peppers, chopped
2 cups chopped hot peppers (or less)
1 quart vinegar
4 cups sugar
¼ cup salt
1 tablespoon whole cloves

Combine all the ingredients in a large preserving kettle. Heat to boiling then boil slowly, stirring occasionally, about 3 hours or until thick. Ladle the relish into clean, hot pint jars to within ½ inch of each jar's top. Wipe off the tops and threads of the jars. Put on prepared lids and seal as the manufacturer directs. Process in a boiling water bath for 15 minutes.

POTTSFIELD PICKLES STONEHOUSE

Sweet, pungent, hot and spicy — this relish is a prize-winner from Dorothy Stonehouse of Presque Isle, Maine. If you have a 1-quart measure, remember that 3 pints = 1½ quarts. The recipe makes 9 (1-pint) jars.

3 pints chopped green tomatoes (2 to 2½ pounds)
3 pints chopped ripe tomatoes (2 to 2½ pounds)
3 pints chopped cabbage (about 2 pounds)
1 bunch celery, chopped
2 sweet red peppers, chopped
1 hot pepper, chopped
3 medium onions, chopped
½ cup pure granulated salt
3 pints vinegar
6 cups sugar
½ cup mustard seed
1½ teaspoons ground cinnamon
½ teaspoon ground cloves

Combine the tomatoes, cabbage, celery, peppers and onions in large mixing bowl, crock or plastic dishpan. Sprinkle with salt and let them stand overnight. Drain well. Combine all the remaining ingredients in a large preserving kettle. Add the drained vegetables and simmer 1 to 2 hours or until the vegetables are very tender. Ladle them into clean hot jars to within ½ inch of each top. Wipe off the tops and threads of the jars with a damp cloth. Put on prepared lids and seal as the manufacturer directs. Process in a boiling water bath for 10 minutes.

HOT DOG RELISH LOPP

Here is a good way to use up zucchini. Jane Lopp of Kalispell, Montana sent us her tangy recipe. She won a blue ribbon at the Northwest Montana Fair with her zucchini relish for hot dogs. The recipe makes about 6 (1 pint) jars.

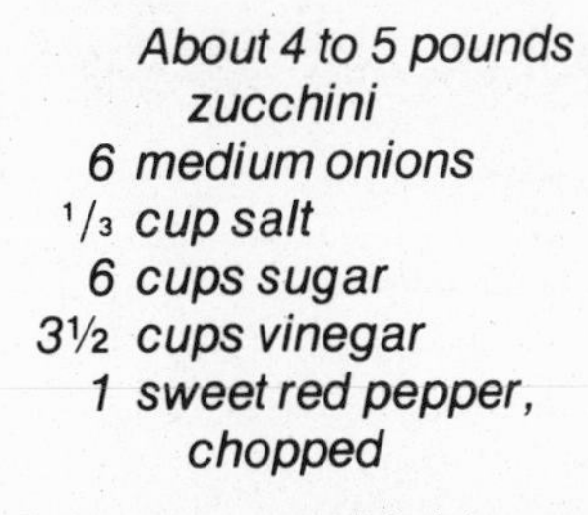

About 4 to 5 pounds zucchini
6 medium onions
1/3 cup salt
6 cups sugar
3½ cups vinegar
1 sweet red pepper, chopped
1 green pepper, chopped
1 tablespoon cornstarch
1 tablespoon turmeric
1 tablespoon nutmeg
1 tablespoon dry mustard
2 teaspoons celery seed

Grind the zucchini (you should have about 10 cups); grind the onion. Mix the zucchini and onions with the salt and let them stand several hours or overnight. Rinse well with cold water and drain. In a large preserving kettle, combine all the remaining ingredients. Add the zucchini and onion; heat to boiling. Reduce heat and simmer 30 minutes. Pack the relish into clean, hot pint jars to within ½ inch of each jar's top. Wipe off the tops and threads of the jars. Put on prepared lids and seal. Process in a boiling water bath for 10 minutes.

RHUBARB RELISH MACDOUGAL

Mrs. Ronald MacDougal of Easton, Maine, serves her blue ribbon sweet-sour relish with meats. You might want to try it with turkey or duck, or spread a spoonful or two on cold roast pork in a sandwich. The recipe makes about 4 (1-pint) jars.

About 2½ to 3 pounds rhubarb
4 medium onions, chopped
3 pounds brown sugar
1 quart vinegar
3 cinnamon sticks
1 teaspoon whole allspice
1 teaspoon salt

Trim the rhubarb and chop finely; you should have about 2 quarts. Tie the spices in a cheesecloth bag. Combine the rhubarb with all remaining ingredients in large preserving kettle. Add the spice bag. Heat until it boils, then reduce the heat and simmer, stirring occasionally to prevent sticking, for about 1 hour or until thick. Ladle the relish into clean hot jars to within ½ inch of each jar's top. Wipe off the tops and threads of the jars with a damp cloth. Put on prepared lids and seal as the manufacturer directs. Process in a boiling water bath for 5 minutes.

SPICED TOMATO PINEAPPLE RELISH BESICH

Mary J. Besich of Black Eagle, Montana, puts the fruit of the vine (tomato, that is) to good use. Her tantalizing relish is almost a chutney — great with meat, over fish, alongside poultry. Try mixing a few spoonfuls with mayonnaise for a marvelous salad dressing. The recipe makes about 9 (½-pint) jars.

About 2 pounds ripe tomatoes
*1 package powdered fruit pectin**
1 can (13½ ounces) crushed pineapple
¼ cup vinegar
½ teaspoon ground cinnamon
2 teaspoons Worchestershire sauce
¼ teaspoon ground cloves
½ teaspoon ground allspice
5 cups sugar

Wash and peel the tomatoes. Chop them coarsely, measure 1 quart and put it into a large saucepan. Heat to boiling, then reduce the heat and simmer 10 minutes, stirring occasionally to prevent sticking. Stir in the pectin, pineapple, vinegar and spices and heat to a full rolling boil. Stir in the sugar and heat again to a full rolling boil; boil 1 minute, stirring constantly. Remove the saucepan from the heat and stir and skim for 5 minutes to keep the fruit from floating. Ladle into clean, hot ½-pint jars to within ½ inch of each jar's top. Wipe off the tops and threads of the jars with a damp cloth. Put on prepared lids and seal as the manufacturer directs. Process in boiling water bath for 5 minutes.

**Follow the pectin package directions for exact boiling time. Some brands require 2 minutes.*

SWEET AND SOUR RELISH MARSHALL

Tasty on hot dogs, burgers, in a grilled cheese sandwich, or with pot roast, this relish won a blue ribbon at the Passaic County 4-H Fair for Becky Marshall from Wayne, New Jersey. The recipe makes about 7 (1-pint) jars.

9 green peppers, seeded
9 sweet red peppers, seeded
9 medium onions, peeled and cut in chunks
9 stalks celery, cut in chunks
Boiling water
3 cups vinegar
2½ cups sugar
1 tablespoon salt
½ teaspoon mustard seed

Grind the peppers, onions and celery, using the coarse blade of a food grinder. Pour enough boiling water over the vegetables to cover them, then let them stand for 10 minutes. Drain well. Combine all the remaining ingredients in a large preserving kettle. Add the vegetables and boil for 15 minutes. Ladle into clean, hot pint jars to within ½ inch of each jar's top. Wipe off the tops and threads of the jars with a damp cloth. Put on the prepared lids and seal as the manufacturer directs. Process in a boiling water bath for 10 minutes.

CHILI SAUCE SMITH

Ann Smith, first place winner at Cobb County's North Georgia State Fair, says you can replace the spices given below with 6 tablespoons of mixed pickling spice. Tie whole spices or pickling spice in a muslin bag or square of cheesecloth. The recipe makes about 3 (1-pint) jars.

5 pounds ripe tomatoes, peeled and chopped
1 medium onion, chopped
1 green pepper, chopped
1 cup brown sugar
2 tablespoons salt
2 teaspoons ground ginger
½ teaspoon nutmeg
2 sticks cinnamon
1 tablespoon mustard seed
½ teaspoon cayenne pepper
1 pint vinegar

Combine all the ingredients except the vinegar in a large preserving kettle. Heat to boiling and boil slowly, stirring frequently, for 2 hours. Stir in the vinegar and boil slowly another hour, stirring frequently to prevent sticking. Ladle the sauce into clean, hot pint jars to within ½ inch of each jar's top. Wipe off the tops and threads of the jars with a damp cloth. Put on prepared lids and seal as the manufacturer directs. Process in a boiling water bath for 15 minutes.

SWEET RELISH PHILLIPS

Debra L. Phillips from Elko, Nevada, tells us her prize-winning recipe has been handed down through the years. The recipe helped her win a trip to the National 4-H Congress. Miss Phillips hopes everyone will enjoy it as much as her family does. She serves it with cold meat. The recipe makes about 8 (1-pint) jars.

- *1 large head cabbage, chopped*
- *1 head cauliflower, separated (into small flowerets)*
- *4 green peppers, chopped*
- *2 sweet red peppers, chopped*
- *6 green tomatoes, chopped*
- *2 large onions, chopped*
- *1 cup pure granulated salt*
- *1 quart cider vinegar*
- *2 cups sugar*
- *2 tablespoons mustard seed*
- *1 tablespoon whole allspice*
- *1 tablespoon whole cloves*
- *1 teaspoon turmeric*

Combine the cabbage, cauliflower, peppers, tomatoes and onions in a large bowl, crock or plastic dishpan. Sprinkle them with salt; add cold water to cover and let stand overnight. Tie the four spices in a cheesecloth bag. Drain well. In a large preserving kettle, combine all the remaining ingredients and heat to boiling. Add the drained vegetables and heat again to boiling. Remove the spice bag. Ladle the relish into clean hot, pint jars to within ½ inch of each top. Wipe off the tops and threads of the jars with a damp cloth. Put on prepared lids and seal as the manufacturer directs. Process in a boiling water bath for 10 minutes.

GRANDMA'S MOCK MINCEMEAT

Ellie Hendrick, from Beavercreek, Oregon, won a trip to the National 4-H Congress for her Food Preservation Project. This recipe won a blue ribbon at the Oregon State Fair. The recipe makes 4 (1-pint) jars.

2 cups chopped green tomatoes
2 cups peeled chopped apples
1 orange, ground (peel and all)
3 cups sugar
1 cup raisins
½ cup vinegar
½ cup butter
3 tablespoons flour
1 teaspoon cloves
1 teaspoon cinnamon
½ teaspoon mace
½ teaspoon nutmeg

Combine all the ingredients in a large preserving kettle and heat to boiling. Simmer until thick. Ladle the mincemeat into clean, hot jars to within 1 inch of each jar's top. Wipe off the tops and threads of the jars with a damp cloth. Put on prepared lids and seal as the manufacturer directs. Process in a boiling water bath for 25 minutes.

Jellies, Jams, Preserves, Conserves, Marmalades And Butters

A perfect jelly is as marvelous a creation as any jewel. Jelly sparkles, has brilliance, clarity, and facets that reflect light and color. And, jelly is ever so much better on a biscuit than a diamond!

One jelly recipe, current at the time of the Revolution, calls for strips of jelly to be separately colored green, red, yellow, and blue. The colors were made from spinach, coccineal, saffron, and violet syrup.

Jellies actually started as what we now call gelatin. In the 16th century, jellies were served with a main dish or as part of the dessert. Jellies were made to hold their shape with hartshorn, calves's feet or, later, isinglass. Isinglass was preferred by the Colonists, though calves' feet were used. Isinglass is a very pure gelatin derived from sturgeons' swimming bladders. It was a Russian discovery, brought to Britain by Dutch traders. Jellies sometimes were fruit flavored, but more often they were merely colorful or spicy additions to dinners. In the 17th century, a favorite jelly was water or white wine stiffened with isinglass, sweetened with sugar and flavored with rosewater and spices. Fruit jellies were made with isinglass or by boiling down fruit juice until it would hold a shape by itself. Well-to-do early Americans often included veritable jelly sculptures in their impressive dinners: floating islands, which used stiffened jelly as a background to support molded shapes; or flummery, which was stiffened, spiced cream molded in imaginative forms.

Marmalade seems to have the most claims to seniority in the preserved fruit family. Quince marmalade, highly spiced and cooked in honey or sugar until it was stiff enough to be cut with a knife, dates back to medieval times. As sugar became more available, all kinds of fruit, vegetables and herbs were candied, and marmalades of various fruits were made. Jam was the "instant" version of marmalade and its more formal relatives, preserves and conserves. Fruit was crushed and quickly boiled in a sugar syrup. Jam eventually became the preferred fruit and sugar concoction in the 18th century.

Fruit butter was an early Yankee favorite. Apple butter was usually prepared outdoors in an enormous cauldron, combining cider, chopped and cored apples, sugar and maybe some spice. Then it had to be stirred all day. Fortunately, the spicy fragrance drew reinforcements to help stir the butter with a wooden paddle.

In this chapter, you will find three non-fruit jellies — Mint, Herb, and Wine, as well as fruit jellies, jams, conserves, preserves and marmalades of all kinds. Don't worry, isinglass is not used, we call for bottled or powdered pectin instead.

Basic Equipment

Always try to use the right equipment to make jelly. Skipping a step or using the wrong size pan could be disastrous, or at the least, very messy. While this list of equipment looks long there are only a very few items that probably are not already in your

Basic Equipment for Jelly. Use the right size pan and correct equipment for best results.

Jelly can be poured into jelly glasses and sealed with paraffin; or sealed in canning jars with 2-piece self-sealing lids.

kitchen. Items that you may need to buy are investments in good jelly.

1. **A preserving kettle** is any large, heavy, broad-beamed pot. "Large" means it must hold at least 8 to 10 quarts. Measure how much water yours will hold, just to be sure it is big enough. Jelly needs to boil high and wide, so a broad surface and adequate depth are essential. The kettle should be unchipped enamel, aluminum or stainless steel. Iron, copper or tin will discolor the fruit and may give an off-taste.
2. **Glasses or jars:** Standard, commercially-made glass canning jars or jelly glasses are the safest containers for jelly because they are designed to stand the heat of sterilizing and the boiling hot jelly. They are a one-time expense that should last you most of your jelly-making years. The only annual purchases you will need to make are paraffin for glasses or new self-sealing lids for jars.
   Glass canning jars with two-piece lids are the best containers for jelly since the paraffin may not always seal perfectly. Check the jars, screw bands and lids carefully and discard any imperfect parts. Jars must be free of cracks, chips, or flaws. Screw bands must be in good working condition. Lids must be new. Never use last year's lids and never re-use lids.
   Jelly glasses are straight-sided, small (usually ½-pint) glass containers. The tops are not threaded, as canning jars are. Glasses are often sold with metal or plastic covers that fit over the top protecting the rim, but not sealing the glasses. Melted paraffin is used to seal glasses.
3. **A large pot or water bath canner** is needed to prepare the glasses or jars. If you are going to put jelly into glasses, you will need a large pot (other than preserving kettle) to sterilize the glasses. If you follow our recommendation and use 2-piece self-sealing lids on canning jars and process in a boiling water bath, you will need a water bath canner.
4. **A jelly bag** is used to strain fruit juice for sparkling clean jelly. You can buy a ready-made bag, often with its own stand, wherever canning supplies are sold. Pour boiling water through the bag before using it for the first time. Always wash and rinse the bag thoroughly after each use. Instructions for making a jelly bag from cheesecloth follow the Basic Equipment.
5. **A candy thermometer** eliminates all guesswork when making jelly without added pectin. You will not need a

thermometer if you add commercial pectin. Select a thermometer with a clamp so you can fasten it to the edge of the kettle. If you live high above sea level, boil some water and make a note of the boiling temperature. By adding 8°F to that figure you will know the proper temperature.

6. **Spoons.** You will need several spoons: a large, long-handled wooden spoon is excellent for comfortable, silent stirring; a slotted metal spoon is best for skimming foam off jelly since the foam stays on the spoon while the jelly slips back to the kettle; use a metal tablespoon or mixing spoon for testing the jelly. You will also need a full set (¼, ½, 1 teaspoon and 1 tablespoon) of measuring spoons.
7. **Knives.** You will need a paring knife and a chopping knife or chef's knife.
8. **Measuring equipment,** properly used, is important for successful jelly. A set of dry (metal) measuring cups (½, ⅓ and 1 cup) and 1-cup and 1-quart liquid (glass) measures are needed. There is a new plastic 1-cup liquid measure available that also measures in metric amounts. It holds 250 milliliters.
9. **A household scale** that can weigh from 4 ounces (¼ pound) up to 10 or 25 pounds is handy for measuring.
10. **A lemon squeezer** or juicer and grater should be at hand, unless you are using bottled lemon juice.
11. **A strainer, food mill, blender, food chopper or grinder** helps prepare fruit. The recipe will tell you if one of these is needed.
12. **A timer** and/or clock with a second hand is also important. Even half a minute of overcooking can scorch jellies, so watch the clock and stay right with your kettle!
13. **A wide-mouth funnel** helps you pour the jelly in the jars, not all over the counter. Some funnels have a measured base, to help you see head space at a glance.
14. **A ladle,** especially one with a little lip or spout, is invaluable for transferring hot jelly from the kettle to the jar.
15. **Long-handled tongs or special jar lifters** should be used for safely handling the hot jars or glasses.
16. **Hot pads, oven mitts, wire cooling racks or folded dish towels** protect your hands and countertops from hot jars.

How to Make a Cheesecloth Jelly Bag

You can make your own jelly bag from a few yards of cheesecloth. Cheesecloth is usually sold in the yard goods section of a store. Cut several 24-inch lengths and put them in a pan. Pour boiling water over the cheesecloth. When cool, lift the pieces out and wring dry. Line a large colander with the pieces, one on top of another. Set the colander in a large bowl; pour in the fruit. Gather together the ends of the cheesecloth and tie them in a large, loose knot. Hang the fruit over a bowl to catch the juice. A cupboard door handle, faucet in the sink, or a broomstick across two chairs make good handles for hanging your homemade jelly bag.

Sealing Jars with Two-piece Lids

There are two methods for finishing jelly in jars with two-piece lids. You may seal the jelly in sterilized jars, or seal the jelly in clean hot jars and process it in a boiling water bath for 10 minutes. More and more canning authorities are recommending the boiling water bath for every canner product — including jellies, jams, pickles, relishes. The processing is an extra guarantee of safety and quality.

Sterilizing Jars

To sterilize: Wash the jars and covers in hot suds and rinse. Put a folded clean towel or wire rack in the bottom of large kettle or water bath canner, then put in the jars. (The towel or rack keeps the jars from bumping and breaking.) Pour in enough hot water to cover the jars. Heat to boiling; cover the pan and boil 10 minutes. Keep the jars in hot water in the pan until ready to fill; lift them out with tongs and drain.

Sealing Sterilized Jars

Keep the jars hot. Prepare the lids as the manufacturer directs. Pour the hot jelly into the jars, put the lids in place, protecting your hands with pot holders or mitts and firmly screw on the bands. Immediately tip each jar over so the hot jelly touches the lid to heat and sterilize it. Tip the jar back upright and set aside to cool in a draft-free place.

Sealing and Processing Jars in a Boiling Water Bath

Wash the jars and screw bands in hot suds, rinse and keep them hot. Prepare the lids as the manufacturer directs. Fill hot, clean jars with hot jelly to within ⅛ inch of the top. Put the lids in place. Protect hands with pot holders or mitts and firmly screw on the bands. Put the jars in a boiling water bath.

Boiling Water Bath for Jelly

Because you are working with ½-pint or 12-ounce jars you will not need a big water bath canner for processing. A large saucepan, Dutch oven or other kettle will do, if you have a cake or other rack that will fit in the bottom, and if the pan is deep enough to hold water to cover the jars on the rack and still has room to boil. Depending on the height of the rack, this would require a pan 7 to 8 inches deep. Put the rack in the bottom of the pan, pour in enough water to almost half full and heat to boiling. At the same time heat more water in a teakettle or other pan. Gently lower the filled and sealed jars onto the rack, then pour in more boiling water, enough to cover. Heat to boiling, then cover and begin timing as the recipe directs.

Sealing Glasses with Paraffin

Always melt paraffin over hot water. Paraffin is flammable and can catch fire if it is heated over direct heat. Use an old double boiler or a coffee can or other expendable container set in another pan of water. Bend the coffee can between your

Melt paraffin in can in pan of simmering water, or old double boiler.

Pour on just enough paraffin to make a layer 1/8 inch thick.

Puncture any air bubbles in the paraffin with tip of a knife.

hands to form a small spout for easier pouring. Break blocks of paraffin into the can or top of the double boiler and heat over water with medium-low heat until melted.

Sterilize the glasses as in Sterilizing Jars. Pour the hot jelly into the hot sterilized glasses set on a level surface. Slowly and carefully pour about 1 tablespoon melted paraffin over the top of each glass of jelly or pour until the paraffin is about ⅛-inch thick. Protect your hands with a pot holder or mitt and gently tilt and tip the glass so the paraffin seals to edge of glass. Break any bubbles in the surface of the paraffin layer by poking them with the tip of a knife.

You may find it easier to dip melted paraffin with a small measuring cup or other small dipper and then pour it onto the jelly.

Set the paraffin-sealed glasses aside in a draft-free place and let them stand overnight. Then cover them with metal covers, waxed paper or plastic wrap.

Basic Ingredients

It takes only four ingredients to make jelly: pectin, fruit juice, sugar and an acid (from lemon juice or tart fruits). Combining these four in just the right amounts, at just the right temperature and for just the right amount of time brings about a remarkable reaction: the fruit juice can hold a shape; a quivering delicate shape that is easy to cut and spread.

We recommend making jelly with commercial pectin according to our recipes. You will find that cooking times are shorter than with natural pectin. There is less chance for error, and you can be sure the jelly will jell when you use commercial pectin and cook the jelly for the exact time given in each recipe. When you have acquired some jelly-making skill and confidence, then you can go on to jelly-making without added pectin. Instructions for both methods are included in the recipes.

1. **Pectin** is what makes jelly jell. Pectin is a natural substance found in many fruits. Apples, cranberries, blackberries, gooseberries, currants, citrus fruits and Concord grapes all have enough of their own pectin that you do not need to add any more to make them jell. Commercially made pectin from apples and citrus fruit can be purchased to add to fruits with low pectin content —

Jelly can be made with fresh fruits, frozen fruits or juice concentrates, canned or bottled fruit juices.

strawberries, blueberries, peaches, apricots, cherries, pineapple and rhubarb.

Commercial pectin comes in two forms:

liquid, in 6 ounce brown bottles

powdered, in 1¾ ounce packages.

Because powdered pectin goes into the kettle with the juice before heating, and liquid pectin goes in after the juice and sugar have boiled, you cannot use them interchangeably. Follow the recipe and use the type it specifies. Always check the date on the pectin box or bottle. Pectin does not hold its strength from one year to the next. This means you should not use any you have left over from last year. Liquid pectin must flow. If it has jelled in the bottle do not use it.

2. **Fruit juice.** Fresh fruits are an obvious choice for jelly-

making, but frozen juice concentrates, unsweetened frozen or canned fruit and juices can become jellies, too. Plan to add pectin to any frozen or canned fruit. Dried fruit can become jelly, also. Cook dried fruit in a small amount of water until tender, then extract juice and make jelly as a recipe for that same fruit directs.

3. **Sugar.** Sugar is sugar, whether it comes from cane or beet. Use granulated sugar, not confectioners' or brown sugar. Honey can be used to replace part of the amount of sugar called for, but not more than half the total amount in any recipe that has no pectin added. If you are adding pectin you can substitute 2 cups of honey for 2 cups of the total amount of sugar called for. Recipes that make 5 to 6 glasses of jelly can be altered, replacing ¾ to 1 cup of sugar with honey. Always use light, mild-flavored honey. Dark, strong honey will overpower the flavor of the fruit.

   Light corn syrup may be used in place of ¼ of the total amount of sugar in recipes without added pectin. If you use powdered pectin you may substitute corn syrup for ½ the amount of sugar. If you use liquid pectin, substitute corn syrup for up to 2 cups of the total amount of sugar.

Step 5: Prepare the fruit as the recipe directs. Here apples are cored and sliced before simmering.

4. **Acid.** Lemon juice is the most common acid in jelly recipes. Use bottled lemon juice for consistent acid content. White cider vinegar is used in the recipe for Herb jelly.

Jelly	**Replacement**	**Proportion**
No added pectin	*Corn syrup*	*1 part syrup to 4 parts sugar*
	Honey	*1 part honey to 2 parts sugar*
Added pectin (powdered or liquid)	*Honey*	*up to 2 cups of total amount of sugar*
Added powdered pectin	*Corn syrup*	*1 part syrup to 1 part sugar*
Added liquid pectin	*Corn syrup*	*1 part syrup to 2 parts sugar*
Jams, Marmalades, Conserves		
	Corn syrup and Honey	*1 part syrup or honey to 2 parts sugar*

Basic Steps With Added Pectin

1. Set out and wash all equipment, wash all working surfaces and hands.
2. Select fruit that is perfect — no bruises, blemishes or bad spots. Size and shape is not important, in fact making jelly is a good way to use oddly sized or shaped fruits. Fruit should be fully ripe for the best flavor.
3. Wash fruit carefully and thoroughly in cold water. Fill the sink, then gently put the fruit in and swish it around. Berries get velvet glove treatment! Always lift fruit out of water and put it in a colander, sieve or on several thicknesses of paper towels to drain. Never leave fruit in the sink and let the water drain out — that just deposits dirt back on the fruit. Do not let the fruit stand too long in water.
4. Hull, stem, peel or pit the fruit as the recipe directs.
5. Prepare the fruit as the recipe directs to extract juice. This may be as simple as mashing (berries), heating with just a little water (grapes), or simmering with a small amount of water until tender (apples).

Step 6: Pour fruit into jelly bag or several layers of cheesecloth in a colander, as shown.

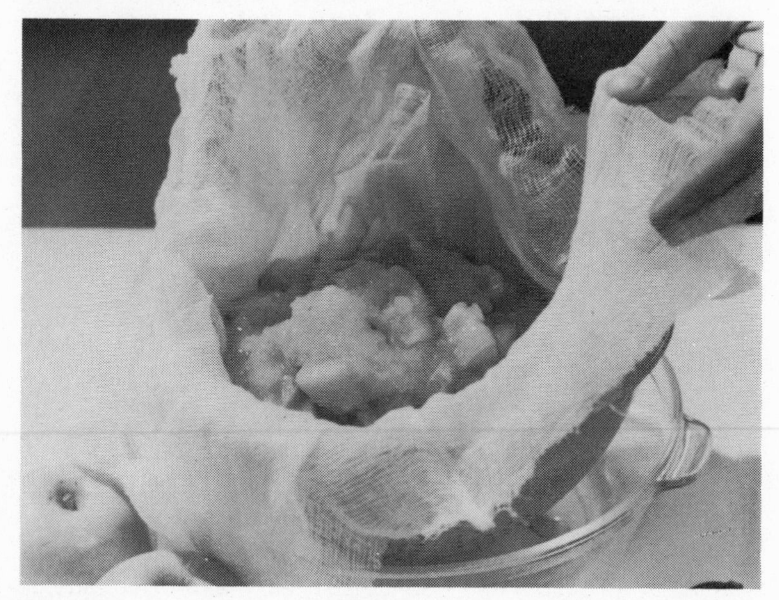

Step 6: Lift corners of the cheesecloth; gather together and tie to form a bag.

Step 6: Hang bag from cupboard door handle or faucet and let the juice drip into a bowl.

Step 7: If using jelly glasses, sterilize them by boiling 15 minutes; melt paraffin.

Step 7: If using canning jars, sterilize them 15 minutes; prepare lids as manufacturer directs.

Step 8: Measure juice and pour amount given in recipe into a preserving kettle. Cook small batches.

6. Prepare your jelly bag on a stand over a bowl or in a colander over a bowl. Pour the fruit into the jelly bag. If you use a cheesecloth bag, tie and hang it over a bowl from a cupboard door handle, faucet or broomstick across two chairs. Let it hang until all the juice has dripped through. The time this takes will vary with the fruit, anywhere from an hour to overnight. If you are in a hurry, you can squeeze or press the jelly bag, but you should then strain the juice one more time through a double thickness of cheesecloth. Pressing pushes sediment through and it can cloud the jelly. Patience is rewarded by clear jelly.
7. While the juice is straining, sterilize jars or glasses (or wash jars, if you are going to process jelly in a boiling water bath). Prepare the lids and melt the paraffin. Keep the jars hot in hot water, a low oven or in dishwasher set on dry.
8. Measure the juice and pour the amount given in the recipe into the preserving kettle. DO NOT DOUBLE THE RECIPE and DO NOT COOK MORE THAN 4 to 6 CUPS OF JUICE AT ONE TIME. Bigger batches of jelly will probably boil over and are less likely to set.
9. Cook as the recipe directs.
10. If using powdered pectin, add it to the juice in the preserving kettle; turn up the heat and bring it to a full rolling boil, stirring constantly. Heat and stir until the boil cannot be stirred down. Add the sugar and heat and stir again to a full boil. Start timing and boil hard for 1 minute.
11. If using bottled pectin, stir sugar into the juice in the preserving kettle; turn up the heat and bring to full rolling boil, stirring constantly. Heat and stir until the boil cannot be stirred down. Stir in bottled pectin; heat and stir again to full rolling boil. Start timing and boil hard for 1 minute.
12. Immediately remove the kettle from the heat.
13. Quickly skim the foam from the jelly with slotted spoon. Put the foam in a small bowl or cup to discard when you have time later.
14. Put a wide-mouth funnel in a hot glass or jar and ladle in the hot jelly. Fill glasses to within ½ inch of the tops; fill jars to within ⅛ inch of the tops.
15. Wipe the rims of the glasses and rims and threads of the jars with a clean, damp cloth, protecting the hand that holds the jar with a pot holder or mitt.
16. Seal glasses with paraffin. Seal jars with self-sealing two-piece lids and process in boiling water bath for 10 minutes.

Step 10: If using powdered pectin, add it to juice in kettle before cooking. Read pectin label directions.

Step 11: If using bottled pectin, add sugar to juice in kettle.

Step 11: Heat juice to full rolling boil, then add bottled pectin to the juice.

Step 13: Immediately skim off foam with a slotted spoon.

Step 14: Using a wide-mouth funnel, ladle jelly into hot glass or jar to level stated in recipe.

17. Set sealed glasses or jars out of the way in a draft-free place and let them stand overnight.
18. Check the seal on paraffin by looking for leaks. On jars, test seal by pushing the center of the lid with your forefinger. If the lid does not give, the jar is sealed.
19. Cover paraffin-sealed glasses with metal covers, waxed paper or plastic wrap. Remove screwbands from jars so they will not rust in place.
20. Wipe the outside of glasses or jars with a clean damp cloth.
21. Label with the type of jelly and date.
22. Store in cool, dark, dry place.

Step 15: Wipe the rims of the glasses and rims and threads of the jars with a damp cloth.

Step 21: After wiping outsides of glasses or jars, label with type of jelly and date.

Basic Steps Without Added Pectin

1. Set out and wash all equipment, wash all working surfaces and hands.
2. When you select fruit, ¾ of it should be ripe, ¼ of it should be underripe. Underripe fruit has more pectin.
3. Wash fruit carefully and thoroughly in cold water. Fill the sink, then put the fruit in and gently swish it around. Be very careful with berries. Lift the fruit out of the water and drain it in a colander, sieve or on paper towels.
4. Cut away any damaged parts of the fruit. Remove the stem and blossom ends but do not peel or core. There is lots of pectin in the peel and core.
5. Prepare the fruit as the recipe directs to extract juice.
6. Prepare your jelly bag on a stand over a bowl or in a colander over a bowl. Pour the fruit into the jelly bag. If you use a cheesecloth bag, tie and hang it over a bowl.
7. Test cooled, cooked juice for pectin strength by mixing about 1 tablespoon juice with 1 tablespoon rubbing alcohol. Stir to mix. If juice has enough pectin the mixture will jell. If juice is too low in pectin it will only jell in small particles and you must add pectin as in jelly made with added pectin. DO NOT TASTE THE ALCOHOL-JUICE MIXTURE. Throw it out and thoroughly wash the utensils used for this test.
8. If you are not using a water bath canner, sterilize the jars or glasses and melt the paraffin. If you are using a water bath canner, prepare the lids. Keep the jars and glasses hot in hot water, a low oven or in the dishwasher set on its dry cycle.
9. Measure the juice into the preserving kettle.
10. Add sugar and lemon juice as the recipe directs; mix well.
11. Place the kettle over high heat; cook and stir until the sugar dissolves, then boil rapidly until the jelly tests done. (Testing Jelly follows the Basic Steps.)
12. Immediately remove from heat.
13. Quickly skim the foam from the jelly with a slotted spoon. Put the foam in small bowl or cup to discard when you have time later.
14. Put a wide-mouth funnel in the hot glass or jar and ladle in the hot jelly. Fill glasses to within ½ inch of the tops. Fill jars to within ⅛ inch of the tops.
15. Proceed as in steps 13 through 22 of Basic Steps for Jelly With Added Pectin.

Testing Jelly

There are three ways to check to see if jelly made without added pectin will jell.

Temperature. Clip a candy thermometer onto the side of the preserving kettle and adjust it so the tip is halfway down in the juice, and the thermometer is completely covered by juice, but not touching the bottom of the pan. The jelly is done when the thermometer reaches 8°F above boiling. If you live at sea level, this is 220°F. If you live several thousand feet above sea level, boil some water, check the boiling temperature with the thermometer and then add 8°F to determine the right temperature for jelly.

Sheet test. After the jelly has boiled for a few minutes, begin making this test. Dip a cool metal spoon into the boiling juice and fill it. Lift the spoon up above the kettle, high enough that it is out of the steam so you can see it. Tilt the spoon slowly so that jelly runs off the side. When jelly is cooked enough it flows together and comes off the spoon in a sheet or flake.

Temperature Test. Clip a candy thermometer to side of kettle. Jelly is done at 8° F above boiling.

Sheet Test. Jelly is done when mixture slides off spoon in a large drop that forms a sheet.

Refrigerator Test. Take kettle off heat. Spoon juice onto cool saucer; place in freezer to see if it jells.

Refrigerator test. Always take the kettle off the heat while making this test. After jelly has boiled for several minutes, spoon a tablespoon or so of juice onto a cool saucer or plate. Put the plate in the freezer for a few minutes. If the mixture jells, the jelly is done.

Nobody's Perfect

There are so many factors involved in jelly and jam making — time, temperature, humidity, juiciness of fruits, proportion of ingredients — that sometimes things go wrong. Here are some common problems, and causes.

1. **Jelly is soft** and does not set. This is often the result of too much sugar, too big a batch, too much juice in the fruit, or not enough acid. You may be able to salvage the jelly by using it as syrup for pancakes, sundaes, sweetening, or by cooking it again to see if it will set.

   If made with powdered pectin, measure the jelly to be recooked. For each 4 cups of jelly, measure ¼ cup sugar, ¼ cup water and 4 teaspoons powdered pectin. Mix the pectin and water; heat and stir until boiling. Add the jelly and sugar and mix well. Bring to a full rolling boil, stirring constantly. Boil hard for 30 seconds. Remove from the heat, skim foam, ladle into hot jars or glasses and seal.

   If made with liquid pectin, measure the jelly to be recooked. For each 4 cups of jelly, measure ¾ cup sugar, 2 tablespoons lemon juice and 2 tablespoons liquid pectin. Heat the jelly to boiling, quickly add the sugar, lemon juice and pectin and bring to full rolling boil. Boil hard for 1 minute. Remove from heat, skim foam, ladle into hot jars or glasses and seal.

   If made without pectin, heat the jelly to boiling for 2 or 3 minutes before testing it with one of the three jelly testing methods. When the jelly test is satisfactory, remove from the heat, skim foam, ladle into hot jars or glasses and seal.
2. **Cloudy jelly** is usually the result of improper straining or squeezing the jelly bag, letting the jelly stand before pouring it into jars, or pouring it too slowly. Overly green fruit can cause the jelly to set too fast and become cloudy.
3. **Crystals** in the jelly indicate too much sugar, or too little, too slow or too long cooking. Follow each recipe exactly

for the amounts of sugar and cooking time. Crystals that form on the top of uncovered jelly are the result of evaporation. Always cover and refrigerate jellies once they have been opened. Some grape jelly may have tartrate crystals. The recipe for Concord Grape Jelly tells you how to prevent these.

4. **Syrupy jelly** comes from too much or too little sugar, or too little acid or pectin.
5. **Stiff or gummy jelly** is the result of overcooking. Stiff jelly may result from too much pectin, too.
6. **Tough jelly** tells you that you used too little sugar.
7. **Weepy jelly** that separates had too much acid added, too thick a layer of paraffin put on top, or was stored in too warm a place. Warm storage and an imperfect seal can also cause jelly and jam to darken at the top of the jar.
8. **Faded jelly.** Hot or too long storage can fade red jellies and jams.
9. **Jelly that ferments or molds** has not been sealed properly. Too little sugar can also cause fermentation.
10. **Floating fruit.** Fruit in jam or preserves floats in jars because it was not ripe enough, was not ground or crushed enough, or was not packed properly.

Recipes

RASPBERRY OR STRAWBERRY JELLY

Add sugar to match your taste: 2½ cups gives a slightly tart jelly. This is one jelly recipe you can double if you have a large preserving kettle. The recipe makes about 4 (½-pint) glasses or jars.

Ingredients

3 pints ripe red raspberries or 1½ quarts ripe strawberries
2 tablespoons lemon juice (for Strawberry Jelly only)
2½ to 3½ cups sugar
½ bottle liquid pectin

Equipment

4(½-pint) jelly glasses or jars with 2-piece self-sealing lids
Paraffin (for glasses)
Colander
Masher
Bowl
Jelly bag
Measuring cups
Preserving kettle
Wooden spoon
Ladle
Wide-mouth funnel

1. Sterilize glasses or wash and rinse jars; keep them hot.
2. If using jars, prepare the lids as manufacturer directs. If using glasses, melt the paraffin.
3. Wash and drain the berries. Mash them with a masher or back of a wooden spoon.
4. Turn the fruit into a jelly bag and let it drain over a bowl or measuring cup several hours or overnight, until you have 2 cups juice. For Strawberry Jelly, use 1⅞ cups of juice (2 cups minus 2 tablespoons) and 2 tablespoons lemon juice.
5. Combine the juice and sugar in a preserving kettle, mix well and heat to a full rolling boil, stirring constantly.
6. Stir in the pectin, return to boil and boil hard for 1 minute, stirring constantly.
7. Immediately remove it from the heat. Skim the foam.
8. Quickly ladle through the funnel into hot glasses or jars. Fill glasses to within ½ inch of each top. Fill jars to within ⅛ inch of each top.
9. Wipe the tops of glasses or jars, inside edge of glasses, and threads of jars with a damp cloth.
10. Seal glasses with ⅛-inch layer of paraffin.
    If using jars, put on the lids and screw bands as the manufacturer directs. Process the jars in a boiling water bath for 10 minutes.

GRAPEFRUIT WINE JELLY

Serve this shimmering jelly with fish, duck, goose, or even roast beef. The recipe makes 5 (½-pint) glasses or jars.

Ingredients

1 cup strained fresh grapefruit juice
1 cup claret or ruby red port wine
3½ cups sugar
½ bottle liquid pectin

Equipment

5 (½-pint) glasses or jars with 2-piece self-sealing lids
Paraffin (for glasses)
Knife
Citrus juicer
Strainer
Measuring cups
3- or 4-quart heavy saucepan
Wooden spoon
Slotted spoon
Ladle
Wide-mouth funnel

1. Sterilize glasses, or wash and rinse jars; keep them hot.
2. If using jars, prepare the lids as the manufacturer directs. If using glasses, melt the paraffin.
3. In a saucepan combine the grapefruit juice, wine and sugar.
4. Heat and stir over very low heat until well blended.
5. Continue to heat over low heat just until the sugar dissolves and bubbles appear at edge of the pan.
6. Immediately remove from heat and stir in the pectin. Skim the foam, if necessary.
7. Ladle through the funnel into hot glasses or jars. Fill glasses to within ½ inch of each top. Fill jars to within ⅛ inch of each top.
8. Wipe the tops of glasses or jars, inside edge of glasses and threads of jars with a damp cloth.
9. Seal glasses with ⅛-inch layer of paraffin.
   If using jars, put on the lids and screw bands as the manufacturer directs.
10. Process the jars in a boiling water bath 10 minutes.

HERB JELLY

Pick your favorite herb or combination of herbs for this jelly. Herb jelly is delicious with meat. This recipe makes 4 (½-pint) glasses or jars.

Ingredients

- ¼ *cup dried herbs (sage, thyme, marjoram, rosemary, tarragon or combination)*
- 2½ *cups boiling water*
- 4 *to 4½ cups sugar*
- ¼ *cup cider or wine vinegar*
- *Few drops green food coloring*
- ½ *bottle liquid pectin*

Equipment

- 4 *(½-pint) jelly glasses or jars with 2-piece self-sealing lids*
- *Paraffin*
- *Large saucepan*
- *Measuring cups*
- *Measuring spoons*
- *Strainer*
- *Wooden spoon*
- *Slotted spoon*
- *Ladle*
- *Wide-mouth funnel*

1. Pour boiling water over herbs; cover and let it steep for 15 minutes. Strain and measure 2 cups of the herb "tea" into a large saucepan.
2. Stir in the sugar, vinegar and food coloring. Heat to a full rolling boil, stirring constantly.
3. Stir in the pectin, return to a boil and boil hard for 1 minute, stirring constantly.
4. Immediately remove from the heat. Skim off the foam.
5. Ladle through the funnel into hot glasses or jars. Fill glasses to within ½ inch of each top. Fill jars to within ⅛ inch of their tops.
6. Wipe the tops of glasses or jars, inside edge of glasses, and threads of jars with a damp cloth.
7. Seal glasses with ⅛-inch layer of paraffin.
   If using jars, put on the lids and screw bands as the manufacturer directs.
8. Process the jars in a boiling water bath for 5 minutes.

Jams, Preserves, Conserves, Marmalades And Butters

Jams, preserves, conserves, all kinds of marmalades and fruit butters are first cousins to jelly. Jam is softer than jelly and has crushed or chopped fruit in it. It is often made with added pectin to give it a slight jell.
Preserves are whole fruits or pieces of fruit simmered in syrup until plump and saturated.
Conserves are like jam, but they are often made from more than one fruit with raisins and nuts added.
Marmalade is a jelly with little pieces of fruit floating throughout. Citrus fruits make the most popular marmalades, but other fruits can be included.
Butter is a very thick, sweetened puree of fruit pulp, often flavored with cinnamon or other spices.

These jelly-kin are prepared in much the same way as jelly; the recipes give you the exact procedures. Jams, preserves, conserves, marmalades and butters must be sealed in jars with two-piece lids and processed in a boiling water bath. Only jellies sealed in sterilized glasses or jars with paraffin or two-piece lids are not processed in a water bath canner. Sealing and processing helps prevent mold and fermentation and guarantees long-lasting quality.

GREEN AND WHITE GRAPE JAM

A delightful change from purple grape jelly, this grape jam is elegant enough to serve at tea. If you like unique spices, try a whole cardamom seed in place of the cinnamon. The recipe makes about 10 (½-pint) jars.

Ingredients

2 pounds seedless green grapes
1 bottle (24 ounces) white grape juice
1 stick cinnamon
7 cups sugar
1 bottle liquid pectin

Equipment

10 (½-pint) jars with 2-piece self-sealing lids
Knife
Preserving kettle
Measuring cups
Wooden spoon
Slotted spoon
Ladle
Wide-mouth funnel
Water bath canner

1. Wash and rinse the jars; keep them hot. Prepare the lids as the manufacturer directs.
2. Wash the grapes, stem and halve them into the preserving kettle along with 1 cup of the grape juice and the cinnamon stick.
3. Heat to a boil, then reduce the heat to a simmer. Cover and simmer 15 minutes.
4. Stir in the remaining grape juice and sugar. Heat to a full rolling boil, then boil hard 1 minute, stirring constantly.
5. Immediately remove the jam from the heat and stir in the pectin.
6. Skim and stir for 5 minutes to prevent the fruit from floating. Remove the cinnamon stick.
7. Ladle the jam into hot jars to within ¼-inch of the tops.
8. Wipe the tops and threads of the jars with a damp cloth.
9. Put on the lids and screw bands as the manufacturer directs.
10. Process in a boiling water bath for 10 minutes. See Basic Steps for Canning Fruit 12 through 19 or Boiling Water Bath for Jelly for processing instructions.

BAR LE DUC (French Currant Jam)

Marvelous on muffins, brioche or croissants, Bar Le Duc is also an elegant topper for ice cream, vanilla mousse, cheesecake, or as an accompaniment to game or pork. The recipe makes 6 (½-pint) jars.

Ingredients

3 pounds currants
2¾ to 3 cups sugar
⅓ cup honey

Equipment

6 (½-pint) jars with 2-piece self-sealing lids
Colander
Preserving kettle
Masher or wooden spoon
Measuring cups
Ladle
Wide-mouth funnel
Water bath canner

1. Wash and rinse the jars; keep them hot. Prepare the lids as the manufacturer directs.
2. Wash and drain the currants, then stem. Measure half the currants into the preserving kettle and heat over low heat, pressing with masher or spoon to help release juice. When juicy, stir in the remaining currants.
3. Heat to boiling, stirring constantly. Stir in the sugar. Heat again to boiling, then reduce the heat and simmer, stirring constantly for 5 minutes.
4. Stir in the honey and simmer 5 minutes longer. Remove from heat and ladle into hot jars to within ¼-inch of each top.
5. Wipe the tops and threads of the jars with a damp cloth.
6. Put on the lids and screw bands as the manufacturer directs.
7. Process in a boiling water bath for 10 minutes. See Basic Steps for Canning Fruit 12 through 19 or Boiling Water Bath for Jelly for processing directions.
8. When the jars have sealed and are cooling, invert or shake them gently from time to time to distribute the fruit in the syrup.

NECTARINE RUM JAM

You can vary this simple recipe in several ways: use peaches or apricots in place of nectarines, omit the rum completely, or replace it with orange-flavored liqueur, peach or apricot brandy, or 2 teaspoons vanilla. The recipe makes about 6 (½-pint) jars.

Ingredients

4 cups chopped peeled ripe nectarines (or peaches or apricots)
*1 package powdered fruit pectin**
5 cups sugar
¼ cup rum, liqueur or brandy OR 2 teaspoons vanilla

Equipment

6 (½-pint) jars with 2-piece self-sealing lids
Knife
Preserving kettle
Wooden spoon
Measuring cups
Ladle
Wide-mouth funnel
Water bath canner

1. Wash and rinse the jars; keep them hot. Prepare the lids as the manufacturer directs.
2. Combine the fruit and pectin in the preserving kettle and heat to boiling over high heat, stirring constantly.
3. Add the sugar; cook and stir until it dissolves.
4. Heat to a full rolling boil, then boil hard 1 minute, stirring constantly.
5. Immediately remove the kettle from the heat. Stir and skim for 5 minutes to prevent the fruit from floating. Stir in the rum.
6. Ladle the jam into hot jars to within ¼-inch of the tops.
7. Wipe the tops and threads of the jars with a damp cloth.
8. Put on the lids and screw bands as the manufacturer directs.
9. Process in a boiling water bath for 10 minutes. See Basic Steps for Canning Fruit 12 through 19 or Boiling Water Bath for Jelly for processing instructions.

**Follow the pectin package directions for exact boiling time. Some brands require 2 minutes.*

BERRY-CHERRY PRESERVES

This sweet combination of cherries, strawberries and raspberries is a favorite for hot buttered muffins and toast. You could use dark, sweet cherries in place of tart cherries. The recipe makes 8 to 9 (½-pint) jars.

Ingredients

- *1 quart strawberries*
- *2 pints raspberries*
- *1½ pounds tart red cherries or tart sweet cherries*
- *½ cup lemon juice*
- *1 box powdered fruit pectin**
- *4 to 5 cups sugar (use 4 cups for sweet cherries)*

Equipment

- *8 or 9 (½-pint) jars with 2-piece self-sealing lids*
- *Colander*
- *Masher*
- *Cherry pitter*
- *Preserving kettle*
- *Measuring cups*
- *Wooden spoon*
- *Ladle*
- *Wide-mouth funnel*
- *Water bath canner*

1. Wash and rinse the jars; keep them hot. Prepare the lids as the manufacturer directs.
2. Wash the berries and cherries. Hull the berries; pit the cherries.
3. Mash the fruits in the preserving kettle. Stir in the lemon juice and powdered pectin.
4. Heat to a full rolling boil, stirring constantly.
5. Add the sugar and heat to a full rolling boil. Boil hard 1 minute.
6. Immediately remove the preserves from the heat. Skim the foam, if necessary.
7. Ladle into hot jars to within ¼-inch of each top.
8. Wipe the tops and threads of the jars with a damp cloth.
9. Put on the lids and screw bands as the manufacturer directs.
10. Process in a boiling water bath for 15 minutes. See Basic Steps for Canning Fruit 12 through 19 or Boiling Water Bath for Jelly for processing instructions.

**Follow the pectin package directions for exact boiling time. Some brands require 2 minutes.*

APRICOT PRESERVES

Select apricots that are ripe but still hard, for best shaped preserves. The recipe makes about 4 (½-pint) jars.

Ingredients

2 pounds apricots
4 cups sugar
¼ cup lemon juice
½ teaspoon ground ginger, nutmeg or cloves or vanilla extract (optional)

Equipment

4 (½-pint) jars with 2-piece self-sealing lids
Measuring cups
Preserving kettle
Wooden spoon
Ladle
Wide-mouth funnel
Water bath canner

1. Wash and rinse the jars, drain and keep them hot. Prepare the lids as the manufacturer directs.
2. Wash, drain and scald the apricots. Peel, pit and halve. You should have about 5 cups of apricot halves.
3. In preserving kettle mix apricots, sugar and lemon juice. Cover and let them stand in a cool place 4 or 5 hours.
4. Heat over medium-high heat, stirring gently occasionally until the sugar dissolves.
5. Heat to a boil. Boil, stirring frequently about 30 minutes or until the apricots are clear. Stir in the spice (and the vanilla, if you want).
6. Remove the preserves from the heat and ladle into hot jars to within ¼-inch of the tops.
7. Wipe the tops and threads of the jars with a damp cloth.
8. Put on the lids and screw bands as the manufacturer directs.
9. Process in a boiling water bath for 15 minutes. See Basic Steps for Canning Fruit 12 through 19 or Boiling Water Bath for Jelly for processing instructions.

RHUBARB PINEAPPLE CONSERVE

Try this superb conserve on toasted homemade whole wheat bread or English muffins. You can even use it as a glaze for ham. The recipe makes 4 (½-pint) jars.

Ingredients

4 cups 1-inch slices rhubarb
1½ cups ½-inch chunks fresh pineapple
2 cups sugar
1 cup light corn syrup
1 tablespoon shredded orange peel
¼ cup orange juice
½ cup coarsely chopped Macadamia nuts, pecans or walnuts

Equipment

4 (½-pint) jars with 2-piece self-sealing lids
Knife
Dry and liquid measuring cups
Grater-shredder
Juicer
Wooden spoon
Ladle
Wide-mouth funnel
Water bath canner

1. Wash and rinse the jars; keep them hot. Prepare the lids as the manufacturer directs.
2. Measure all the ingredients except the nuts into the preserving kettle.
3. Heat, stirring constantly, until the mixture is very thick, about 45 minutes. (Remember that the conserve will thicken a little more as it stands.)
5. Ladle into hot jars to within ⅛-inch of the top.
6. Wipe off the tops and threads of the jars with a damp cloth.
7. Put on the lids and screw bands as the manufacturer directs.
8. Process in a boiling water bath for 10 minutes. See Basic Steps for Canning Fruit 12 through 19 for processing directions.

ELEGANT APPLE CONSERVE

Stir a few tablespoons of this into muffin batter before baking for a special treat. Pass more apple conserve to spread on the oven-fresh hot muffins. The recipe makes about 10 (½-pint) jars.

Ingredients

3 pounds ripe apples
5 to 5½ cups sugar (or 4 cups sugar, 1 to 1½ cups honey)
⅓ cup cider vinegar
½ cup raisins or currants
½ cup chopped walnuts or pecans
½ cup red cinnamon candies
1 teaspoon grated lemon peel
2 tablespoons lemon juice
½ bottle liquid pectin

Equipment

10 (½-pint) jars with 2 piece self-sealing lids
Knife
Food grinder or blender
Measuring cups
Measuring spoons
Grater
Lemon juicer
Preserving kettle
Wooden spoon
Slotted spoon
Ladle
Wide-mouth funnel
Water bath canner

1. Wash and rinse the jars; keep them hot. Prepare the lids as the manufacturer directs.
2. Wash the apples, remove the stem and blossom ends and core them. Run the apples through the coarse blade of a food grinder or chop, a few at a time, in a blender. Measure 4 cups chopped apples into the kettle.
3. Stir in all the remaining ingredients except the pectin.
4. Heat to boiling over high heat, stirring constantly. Boil hard 1 minute.
5. Immediately remove from the heat and stir in the pectin.
6. Stir and skim for 5 minutes to prevent fruit from floating.
7. Ladle quickly into hot jars to within ¼-inch of each top.
8. Wipe the tops and threads of the jars with a damp cloth.
9. Put on the lids and screw bands as the manufacturer directs.
10. Process in a boiling water bath for 10 minutes. See basic Steps for Canning Fruit 12 through 19 or Boiling Water Bath for Jelly for processing directions.

FLORIDA CONSERVE

Delicious with cream cheese on brown bread, this conserve can also be spread over a cheesecake to make a party dessert. The recipe makes about 7 (½-pint) jars.

Ingredients

6 oranges
5 cups water
5½ to 6 cups sugar
¼ cup lime juice or vinegar
1 or 2 sticks cinnamon
½ cup raisins
½ cup flaked coconut

Equipment

7 (½-pint) jars with 2-piece self-sealing lids
Knife
Food grinder or blender
Preserving kettle
Wooden spoon
Measuring cups
Candy thermometer or metal spoon
Ladle
Wide-mouth funnel
Boiling water bath

1. Wash and rinse the jars; keep them hot. Prepare the lids as the manufacturer directs.
2. Quarter the oranges and remove the peel to chop; reserve the fruit.
3. Put the peel through the coarse blade of a food grinder or chop in a blender. Put the chopped peel in the preserving kettle with water to cover and boil until tender, about 20 minutes.
4. Meanwhile, cut the center membrane from the fruit and remove any seeds. Dice the fruit, reserving the juice.
5. Add the fruit and juice to the kettle and simmer uncovered about 20 minutes or until reduced by half. You should have about 6 cups.
6. Stir in the sugar, lime juice, cinnamon and raisins; stir until the sugar dissolves.
7. Simmer about 30 minutes until the syrup is thick and the conserve tests done, as for jelly.
8. Immediately remove the conserve from the heat and stir in the coconut.
9. Quickly ladle it into hot jars to within ¼-inch of the top.

10. Wipe off the tops and threads of the jars with a damp cloth.
11. Put on the lids and screw bands as the manufacturer directs.
12. Process in a boiling water bath for 10 minutes. See Basic Steps for Canning Fruit 12 through 19 or Boiling Water Bath for Jelly for processing instructions.

MARVELOUS MIDSUMMER MARMALADE

Peaches, cantaloupe, orange, lemon and ginger — a glorious combination to capture in the summer and serve at the winter meals. The recipe makes about 5 (½-pint) jars.

Ingredients

4 cups diced peeled cantaloupe (about a 2 to 2½ pound melon)
1 orange
1 lemon
3 cups diced, peeled peaches
4½ to 5 cups sugar
½ teaspoon salt
3 tablespoons minced crystallized ginger

Equipment

5 (½-pint) jars with 2-piece self-sealing lids
Knife
Preserving kettle
Wooden spoon
Measuring cup
Measuring spoons
Ladle
Wide-mouth funnel
Water bath canner

1. Wash and rinse the jars; keep them hot. Prepare the lids as the manufacturer directs.
2. Measure the cantaloupe into the preserving kettle.
3. Wash the orange and lemon and finely chop the whole fruit — including the peel. Add the chopped fruit to the cantaloupe.
4. Heat slowly to boiling, then boil 10 minutes, stirring occasionally.
5. Stir in the peaches, sugar and salt and return the fruit to a boil.
6. Boil hard for 20 minutes, stirring often to prevent sticking and scorching.

(Continued On Next Page)

7. Stir in the ginger and boil another 15 minutes, or until as thick as you wish, stirring constantly.
8. Immediately remove the kettle from heat. Stir and skim for 5 minutes to prevent the fruit from floating.
9. Ladle into hot jars to within ¼-inch of each top.
10. Wipe the tops and threads of jars with a damp cloth.
11. Put on the lids and screw bands as the manufacturer directs.
12. Process in a boiling water bath for 10 minutes. See Basic Steps for Canning Fruit 12 through 19 or Boiling Water Bath for Jelly for processing directions.

BLUEBERRY MARMALADE

An all-time favorite for toast, this marmalade can be used in a jelly roll or as a filling for dessert crepes. This recipe makes 7 (½ pint) jars.

Ingredients

1 medium orange
1 lime or lemon
2/3 cup water
2 pints blueberries
5 cups sugar
½ bottle liquid pectin

Equipment

7 (½-pint) jars with 2-piece self-sealing lids
Colander
Grater-shredder
Preserving kettle
Masher or wooden spoon
Knife
Measuring cups
Ladle
Wide-mouth funnel
Water bath canner

1. Wash and rinse the jars; keep them hot. Prepare the lids as the manufacturer directs.
2. Shred or grate only the colored outer portion of the orange and lime peel.
3. Combine the peel and water in a kettle and heat to boiling.
4. Reduce the heat, cover and simmer 10 minutes, stirring occasionally.
5. Meanwhile, wash the blueberries, drain and mash.

6. Cut all the white portion of the peel from the orange and lime and discard. Chop the fruit pulp finely or blend it in a blender. Add the pulp to the cooked peel along with the blueberries.
7. Heat the fruit to boiling, then reduce the heat, cover and simmer about 10 minutes.
8. Stir in the sugar and heat to a full rolling boil, stirring constantly.
9. Stir in the pectin and heat again to a full rolling boil. Boil hard 1 minute, stirring constantly.
10. Immediately remove the kettle from heat. Skim the foam.
11. Ladle into hot jars to within ¼-inch of each top.
12. Wipe off the tops and threads of the jars with a damp cloth.
13. Put on the lids and screw bands as the manufacturer directs.
14. Process in a boiling water bath for 10 minutes. See Basic Steps for Canning Fruit 12 through 19 or Boiling Water Bath for Jelly for processing instructions.

PLUM-PERFECT BUTTER

Watch this tasty butter carefully as it cooks; it is thick and scorches easily. The recipe makes 5 (½-pint) jars.

Ingredients

4 pounds large red or purple plums
½ cup water, or apple, grape or orange juice
4½ to 5 cups sugar

Equipment

5 (½-pint) jars with 2-piece self-sealing lids
Knife
Preserving kettle
Food mill
Sieve or blender
Measuring cups
Wooden spoon
Ladle
Wide-mouth funnel
Water bath canner

1. Wash and rinse the jars; keep them hot. Prepare the lids as the manufacturer directs.
2. Wash, slice and pit the plums.
3. Heat the plums and water or juice in the preserving kettle over medium heat, stirring occasionally, just until plums are soft.
4. Press them through a food mill or sieve, or puree in a blender.
5. Measure the puree — you should have about 6 cups.
6. Combine the puree and sugar in a kettle and simmer over medium heat, stirring frequently until very thick, about 1 to 1½ hours.
7. Remove the puree from the heat and ladle into hot jars to within ¼-inch of each top.
8. Wipe the tops and threads of jars with a damp cloth. Put on the lids and screw bands as manufacturer directs.
9. Process in a boiling water bath for 10 minutes. See Basic Steps for Canning Fruit 12 through 19 or Boiling Water Bath for Jelly for processing instructions.

SPICY PLUM BUTTER: Add 2 sticks of cinnamon and 1 teaspoon whole cloves to the puree along with the sugar. Remove the spices before ladling butter into jars.

SUNSHINE BUTTER

Some dreary day you can brighten your kitchen by putting up a batch of this cheery spread. The recipe makes 6 (½-pint) jars.

Ingredients

- *2 pounds carrots*
- *1 package (8 ounces) or 2 cups dried apricots*
- *1 orange*
- *1 cup water*
- *1 cup light corn syrup*
- *1 cup sugar*
- *1 teaspoon ground cinnamon*
- *¼ teaspoon salt*

Equipment

- *6 (½-pint) jars with 2-piece self-sealing lids*
- *Knife*
- *Measuring cups*
- *Preserving kettle*
- *Food mill or sieve*
- *Wooden spoon*
- *Ladle*
- *Wide-mouth funnel*
- *Water bath canner*

1. Wash and rinse the jars; keep them hot. Prepare the lids as the manufacturer directs.
2. Wash, trim and peel the carrots; cut them in 1-inch chunks. Wash the orange, cut it in eighths and remove any seeds.
3. Combine the carrots, orange pieces and apricots in the preserving kettle along with the water. Cover and heat to boiling. Reduce heat and simmer 15 minutes.
4. Press through a food mill or sieve.
5. Return the puree to the kettle along with all the remaining ingredients; mix well.
6. Heat the puree to boiling, then reduce the heat and simmer 5 minutes, stirring frequently. The butter will be very thick.
7. Remove the butter from the heat and ladle it into hot jars to within ¼-inch of each top.
8. Wipe off the tops and threads of jars with a damp cloth.
9. Put on the lids and screw bands as the manufacturer directs.
10. Process in a boiling water bath for 10 minutes. See Basic Steps for Canning Fruit 12 through 19 or Boiling Water Bath for Jelly for processing instructions.

PENNSYLVANIA DUTCH APPLE BUTTER

The fragrance of cooking apple butter smells like the essence of apple pie. It is likely to draw a crowd. Fortunately this recipe makes enough to share. You can use canned or bottled apple juice or cook 1½ to 2 pounds apples with 1½ cups water about 25 minutes (or until soft). Put the soft apples in a jelly bag and let it drain until you have 2 cups of juice. You can squeeze bag to hurry the juice along, since clarity is not important here, as it is in making jelly. The recipe makes about 12 (½-pint) jars.

Ingredients

8 pounds tart apples
2 cups apple juice or cider
1 cup sugar
1 cup dark corn syrup
2 teaspoons ground cinnamon
½ teaspoon ground cloves

Equipment

12 (½-pint) jars with 2-piece self-sealing lids
Knife
Preserving kettle
Measuring cups
Measuring spoons
Wooden spoon
Food mill or sieve
Ladle
Wide-mouth funnel
Water bath canner

1. Wash and rinse the jars; keep them hot. Prepare the lids as the manufacturer directs.
2. Wash the apples well; remove the stem and blossom ends and cut them in quarters or chunks. Do not peel or core.
3. Put the apples and juice in the preserving kettle. Cover and simmer over very low heat, stirring occasionally, about 1 hour or until apples are very soft.
4. Press the soft apples through a food mill or sieve. Measure the puree.
5. Pour 3½ quarts of puree into the kettle and stir in the remaining ingredients. Leave the kettle uncovered and heat to boiling. Reduce the heat and simmer very slowly about 3 to 5 hours or until very thick, stirring occasionally. During the last hour or so of cooking stir more often to prevent scorching and partially cover the kettle to prevent spattering.

6. Remove the butter from the heat and ladle into hot jars to within 1/4-inch of each top.
7. Wipe the tops and threads of the jars with a damp cloth. Put on lids and screw bands as manufacturer directs.
8. Process in a boiling water bath for 10 minutes. See Basic Steps for Canning Fruit 12 through 19 or Boiling Water Bath for Jelly processing directions.

Freezer Jams

Freezer jams and jellies are uncomplicated and quick — the perfect starter for a beginning food preserver. Because they are unprocessed, they taste like fresh fruit — and, they must be kept in the freezer or refrigerator. In the refrigerator, they keep only for three weeks.

Freezer jams and jellies are simple enough for six-year olds, in fact, and practically goof-proof. All you do is prepare the fruit, mix in sugar and liquid pectin, spoon the mixture into containers and then freeze or refrigerate. You do not even need canning jars and lids or jelly glasses and paraffin. Use small containers (1/2- to 1-pint) that have tight-fitting covers and can be dipped in boiling water.

These jams are great to make when you have lots of fruit but little time or few canning jars. Kids can even put up freezer jams to sell as Scout or club projects.

Basic Equipment

Colander or sieve to drain the fruit
Large mixing bowl
Masher or wooden spoon
Measuring cup
Knife
Lemon squeezer (optional)
Jelly bag (for jelly recipes)
Measuring spoons
Teakettle
Containers, such as small boilable refrigerator dishes, baby food jars, or any pretty container with a tight fitting lid.

Basic Steps

1. Select the containers. Be sure they will hold the total amount of jam the recipe makes. Wash and rinse them with boiling water. Drain and set aside. (Running them through a dishwasher can replace the washing and rinsing.)
2. Wash the fruit thoroughly. Fill the sink with cold water, then gently dump in the fruit. Swish it around with your hands to encourage any dirt to leave. Lift the fruit from the water with your hands and put it in a colander or sieve on a drainboard to drain. Never let the water drain out and then remove the fruit — that just redeposits dirt back on the fruit.
3. Hull, pit or seed fruit as recipe directs, putting the prepared fruit into mixing bowl as you go.
4. Mash fruit with masher or the back of a wooden spoon. Mash thoroughly for thick jam of even consistency; mash slightly less for a chunky jam.
5. Stir in the sugar.
6. Mix the liquid pectin and lemon juice or water. Be sure to use pectin that is fresh — check the date on the bottle. Also tip bottle to be sure the pectin flows. If it has jelled, do not use it.
7. Stir the pectin mixture into the sweetened fruit and mix gently, but thoroughly, for 3 minutes. Do not worry if the sugar does not completely dissolve in that time. Some recipes call for powdered pectin which is heated with water before being added to fruits.
8. Spoon the jam or jelly into containers. Wipe off the rims of the containers with a clean damp cloth.
9. Cover the containers with lids. If you should run short of covers, you can use heavy plastic wrap of foil fastened with a rubber band or string.
10. Set the jam or jelly aside until it is set. This may take a few hours, but might require 24 hours. Tilt container slightly. If the jam or jelly moves, it is not set.
11. When set, put the jam or jelly in the freezer until ready to use, or keep it in the refrigerator for up to 3 weeks. Thaw at room temperature or in the refrigerator. Always refrigerate thawed jam or jelly.

FREEZER RASPBERRY JELLY

Some of this jelly spread on toast or an English muffin can make mornings much more bearable. The recipe makes about 5 cups.

Ingredients

1½ quarts ripe raspberries
5 cups sugar
2 tablespoons water
½ bottle liquid pectin

Equipment

Containers with tight fitting lids
Colander
Large mixing bowl
Masher
Jelly bag
Measuring cups
Measuring spoons
Wooden spoon

1. Wash containers, rinse with boiling water.
2. Wash the berries and drain well.
3. Working with a handful or two at a time, thoroughly crush the berries with a masher or wooden spoon; then put them in a jelly bag. Squeeze out the bag and measure the juice — you will need 2½ cups.
4. Pour the juice into the mixing bowl and stir in the sugar. Let it stand 10 minutes. Combine the water and pectin in a small bowl. Stir into juice and stir 3 minutes (not all the sugar will be dissolved).
5. Quickly pour the jelly into containers and cover with lids.
6. Let the jelly stand at room temperature until set. This may take up to 24 hours.
7. Store the jelly in the freezer.

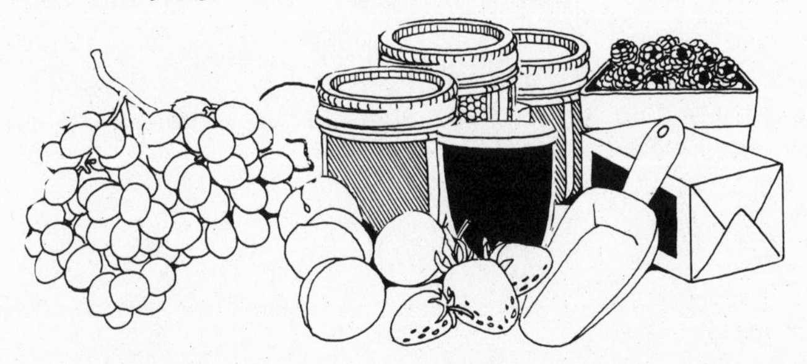

FROZEN CHERRY-BERRY JAM

Berries and cherries make a delightful combination. Raspberries, strawberries or blackberries may be used.

Ingredients

- *1½ pint ripe raspberries or strawberries, or 1 pint blackberries*
- *1¼ pounds tart red cherries*
- *5 cups sugar*
- *¾ cup water*
- *1 box powdered fruit pectin**

Equipment

- *Containers with tight-fitting lids*
- *Colander*
- *Large mixing bowl*
- *Measuring cups*
- *Food grinder or blender*
- *Cherry pitter*
- *Small saucepan*
- *Wooden spoon*

1. Wash the containers, rinse with boiling water.
2. Wash the fruits, drain. Hull the strawberries or crush and measure 1½ cups of berries into a large mixing bowl.
3. Pit the cherries and put them through a grinder or blend in blender. Measure 1½ cups of cherries and add them to the berries.
4. Stir in the sugar.
5. In small saucepan, combine the pectin and water.
6. Heat to a full rolling boil and boil hard 1 minute.
7. Stir the boiled pectin into the fruits and stir 3 minutes (not all the sugar will be dissolved).
8. Pour the jam into containers and cover with lids.
9. Let the jam stand at room temperature until set. This may take up to 24 hours.
10. Store the jam in the freezer.

**Follow the pectin package directions for exact boiling time. Some brands require 2 minutes*

SWEET CHERRY JAM

This recipe uses powdered pectin, so there is a little cooking involved. Thaw the jam to spoon over ice cream for a simple version of Cherries Jubilee. The recipe makes about 3 cups.

Ingredients

2 pounds dark sweet cherries
4 cups sugar
¼ cup lemon juice
*1 package powdered pectin**
1 cup water
2 tablespoons orange-flavored liqueur, kirsch, or brandy (optional)

Equipment

Containers with tight-fitting lids
Colander
Knife
Cherry pitter (optional)
Measuring cups
Saucepan
Large mixing bowl
Wooden spoon

1. Wash the containers, rinse with boiling water.
2. Wash the cherries, drain and stem them. Pit, then quarter, or cut the cherries from pits with knife. You should have about 4 cups.
3. In a mixing bowl, stir together the cherries, sugar and lemon juice. Let them stand ½ hour, stirring occasionally.
4. Meanwhile, in a saucepan combine the pectin and water.
5. Heat to a full rolling boil and boil hard 1 minute.
6. Stir the boiled pectin into cherries along with the liqueur. Stir for 3 minutes.
7. Pour the jam into containers and cover with lids.
8. Let the jam stand at room temperature until cool.
9. Store the jam in the freezer.

**Follow the pectin package directions for exact boiling time. Some brands require 2 minutes.*

ORANGE MARMALADE, FREEZER-STYLE

You do have to cook the orange peel, for better flavor, but this marmalade is still simpler to make than its traditional cousin. The recipe makes 4¾ cups.

Ingredients

2 medium oranges
¾ cup water
4 cups sugar
1/3 cup lemon juice
½ bottle liquid pectin

Equipment

Containers with tight-fitting lids
Knife
Scissors (optional)
Small saucepan
Small and large mixing bowls
Masher
Measuring cups

1. Wash the containers, rinse with boiling water.
2. Wash the oranges. Cut them in quarters and remove the peel.
3. Cut off about half of the white portion of the peel and discard. Cut (or snip with scissors) the colored portion of peel into thin slivers.
4. Put slivered peel and water in the saucepan and heat to boiling.
5. Meanwhile, section the oranges. Discard any seeds and the membrane. In a small mixing bowl, mash the orange sections. Add the peel and liquid. Measure the fruit, peel and liquid and pour 1 2/3 cups of them into the large mixing bowl, add water if you have less than 1 2/3 cups.
6. Stir in the sugar and let the mixture stand 30 minutes.
7. In a small bowl combine the lemon juice and pectin. Stir them into the fruit mixture and stir 3 minutes (not all sugar will be dissolved).
8. Quickly ladle the marmalade into containers and cover them with lids.
9. Let the marmalade stand at room temperature until set. This may take up to 24 hours.
10. Store in the freezer.

Blue Ribbon Recipes

From Maine to Washington, champion jelly and jam makers sent in their treasured recipes. You will find clear, classic jellies and unusual fruit combinations to add to your repertoire. Even city dwellers with small kitchens can put up a small batch of glittering preserves. Be sure to review the basic equipment and basic steps before turning on the heat under your preserving kettle. These jellies, jams, preserves, conserves and butters make excellent gifts, too. Fortunately for us, sugar became a common commodity during the heyday of State fairs — around 1868. Competition and recipes exchanged helped develop the delectable blue ribbon recipes shared in this section.

APPLE JELLY

Both Mrs. H.J. Bigler in Montana and Mrs. Leta M. Meyers in Illinois won prizes for their apple jelly recipes. Mrs. Bigler adds a piece of Kleenex to her jelly for the 1 minute boiling time, then discards it. We think it probably helps to collect the foam. The recipe makes about 10 (½-pint) jars or glasses.

5 pounds firm ripe juicy apples
4 to 5 cups water
*1 package powdered fruit pectin**
8 to 9 cups sugar
Few drops red food coloring (optional)

Cut the blossom and stem ends from the apples; cut the fruit in chunks into a preserving kettle. Add the water and cover and simmer for 10 minutes. Mash with a masher or back of spoon and simmer 5 minutes longer. Pour the pulp into a jelly bag or 4 layers of cheesecloth in a colander. Hang the bag, or tie the ends of the cheesecloth together to form a bag and hang. Let

(Continued On Next Page)

the juice drip into a bowl. Add a few drops red food coloring to the juice, if desired. Measure 7 cups juice into a large preserving kettle. Stir the pectin into the juice and heat to a full rolling boil. At once, add the sugar and heat to full rolling boil, stirring constantly. Boil hard 1 minute. Immediately remove the jelly from the heat. Skim off the foam. Ladle into sterilized jelly glasses to within ½ inch of each top, or ladle into clean, hot ½-pint jars to within ⅛ inch of each top. Wipe off the tops and threads of the jars and the inside rims of the glasses with a damp cloth. Put prepared lids on the jars and seal as the manufacturer directs. Invert jars quickly so hot jelly touches lid, then stand them upright again. Seal the glasses with paraffin.

**Follow the pectin package directions for the exact boiling time. Some brands require 2 minutes.*

PENNY PINCHER'S APPLE BUTTER: Take the pulp left in jelly bag after extracting juice for Apple Jelly and press it through a sieve or food mill. Sweeten to taste with brown sugar, cinnamon and cloves. Simmer it over very low heat until thick. Seal in hot ½-pint jars (it will not make more than 1 or 2 pints) and process in a boiling water bath for 10 minutes.

MINT JELLY BIGLER

Mrs. Bigler's sweepstakes winner at the Utah State Fair is a simple variation of Apple Jelly. To make mint juice, pour 1 cup boiling water over 1 cup fresh mint leaves; squeeze gently and strain. Stir ¼ cup mint juice into the apple juice in the Apple Jelly recipe and proceed as directed. Tint with a few drops green food coloring, if desired.

CHERRY JELLY HAYNES

This sparkling red jelly comes from Myrna Haynes, a long-time resident of Pueblo, Colorado, now living in Montrose. It took a prize at the Colorado State Fair. Your family will award you a blue ribbon when you serve this with muffins, scones, or with peanut butter on graham crackers. The recipe makes about 6 (½-pint) jars or glasses.

2½ -3 pounds firm ripe cherries
½ cup water
*1 package powdered fruit pectin**
4 cups sugar

Wash and stem the cherries, but do not pit them. Crush the berries in a large saucepan or preserving kettle. Add ½ cup water, cover and simmer for 10 minutes. Pour the fruit into a jelly bag or 4 layers of cheesecloth in a colander. Hang the bag, or tie the ends of the cheesecloth together to form bag and hang. Let the juice drip into a bowl. You may squeeze jelly bag from the top, but squeeze gently. Add water if necessary to make 3 cups juice. Stir the pectin into the juice and heat to a full rolling boil. At once, add the sugar and heat again to a full rolling boil, stirring constantly. Boil hard 1 minute. Immediately remove the jelly from the heat. Skim off the foam. Ladle it into sterilized jelly glasses to within ½ inch of each top, or ladle into clean hot ½-pint jars to within ⅛ inch of each top. Wipe off the tops and threads of the jars and the inside rims of the glasses with a damp cloth. Put prepared lids on the jars and seal as the manufacturer directs. Invert jars quickly, so the hot jelly touches the lid, then stand them upright again. Seal the glasses with paraffin.

**Follow the pectin package directions for the exact boiling time. Some brands require 2 minutes.*

CRAB APPLE LEEK

Another prize-winning recipe from Sheila Leek, this clear, lovely jelly helped her win a trip to the National 4-H Congress: The recipe makes about 10 (½-pint) jars or glasses.

5 pounds ripe crab apples
5 cups water
*1 package powdered fruit pectin**
9 cups sugar

Remove the blossom ends and stems from the apples. Cut them in chunks into a preserving kettle. Add water and simmer, covered, for 10 minutes. Pour the fruit into a jelly bag or 4 layers of cheesecloth in a colander. Hang the bag or tie the ends of cheesecloth together to form bag and hang. Let the juice drip into a bowl. Measure 7 cups juice into a large preserving kettle. Stir in the pectin and heat to a full rolling boil. At once, add the sugar and heat to a full rolling boil, stirring constantly. Boil hard 1 minute. Immediately remove the kettle from the heat. Skim off the foam. Ladle the jelly into sterilized jelly glasses or clean, hot jars to within ½ inch of each top. Wipe off the tops and threads of the jars, and the inside rims of the glasses, with a damp cloth. Put prepared lids on the jars and seal as manufacturer directs. Invert the jars quickly so the hot jelly touches the lid, then stand them upright again. Seal the glasses with paraffin.

** Follow the pectin package directions for the exact boiling time. Some brands require 2 minutes.*

CRANBERRY OR CRANBERRY-APPLE JELLY JACKSON

Beautifully sparkling and crimson describes this winner made by Mary Jo Jackson from Apache, Oklahoma. Miss Jackson won a trip to the National 4-H Congress with the help of this recipe. You can make it any season, though it is a natural partner for turkey. The recipe makes about 6 to 7 (½-pint) jars or glasses.

4 cups (1 quart bottle) cranberry juice cocktail or cranberry-apple juice
*1 packaged powdered fruit pectin**
4½ cups sugar

Combine the juice and pectin in a large saucepan or preserving kettle. Heat to a full rolling boil. Add the sugar, heat again to a full rolling boil and boil hard 1 minute, stirring constantly. Remove from the heat immediately and skim off the foam. Ladle the jelly into clean, hot ½-pint jars or sterilized jelly glasses to within ½ inch of each top. Wipe off the tops and threads of the jars, or the tops of the glasses, with a damp cloth. Put prepared lids on the jars and seal as the manufacturer directs; seal glasses with paraffin. Process the jars in a boiling water bath for 5 minutes, or invert quickly so the hot jelly touches the lids, then stand them upright again.

**Follow the pectin package directions for exact boiling time. Some brands require 2 minutes.*

APRICOT PINEAPPLE JAM BIGLER

Two sun-colored fruits combine for marvelous flavor and color in this jam. The recipe won a blue ribbon at the Utah State Fair for Mrs. H.J. Bigler. It makes about 5 (½-pint) jars.

1 pound ripe apricots
1 can (8 ounces) crushed pineapple
2 tablespoons lemon juice
½ package powdered fruit pectin (2 tablespoons)*
4 cups sugar

Wash, pit and grind the apricots. Measure 2¼ cups pulp into a large preserving kettle along with the pineapple, lemon juice and pectin. Heat to a full rolling boil. Stir in the sugar; heat again to boiling and boil hard 1 minute. Immediately remove from the heat. Stir and skim for 5 minutes to prevent floating fruit. Ladle into clean, hot jars to within ½ inch of top of jar. Wipe off the tops and threads of the jars with a damp cloth. Put on the prepared lids and seal as the manufacturer directs. Process in a boiling water bath for 5 minutes.

**Follow the pectin package directions for exact boiling time. Some brands require 2 minutes.*

GRAPE JELLY MEYERS

There is nothing like homemade Concord grape jelly, especially for peanut-butter and jelly sandwiches. Leta M. Meyers, from Chesterfield, Illinois, tells you how it is made. The recipe was a blue ribbon winner at the Macoupin County Fair. The recipe makes about 8 (½-pint) recipes.

3½ to 4 pounds fully-ripe Concord grapes
½ cup water
1 box powdered fruit pectin *
7½ cups sugar

Crush the grapes in a large preserving kettle. Add the water and simmer, covered, 10 minutes. Put the grapes into a jelly bag or 4 layers of cheesecloth in colander. Hang the bag or tie the ends of the cheesecloth together to form a bag and hang. Let the juice drip into a bowl. Measure 5½ to 6 cups juice into a large preserving kettle. Stir in the pectin and heat to a full rolling boil. At once, add the sugar and heat to a full rolling boil, stirring constantly. Boil hard 1 minute. Immediately remove the kettle from the heat. Skim off the foam. Ladle the jelly into sterilized jelly glasses or clean, hot jars to within ½ inch of the tops of the glasses or to within ⅛ inch of each jar's top. Wipe off the tops and threads of the jars, and the inside rims of the glasses, with a damp cloth. Put prepared lids on the jars and seal as the manufacturer directs. Seal the glasses with paraffin. Invert the jars quickly so the hot jelly touches the lid, then stand them upright again.

**Follow the pectin package directions for exact boiling time. Some brands require 2 minutes.*

ORANGE JELLY JOHNSON

Regina Cross Johnson of Commerce City, Colorado, sent us her any-season recipe that took first prize at the Colorado State Fair. It is ideal for last-minute gift-giving. The recipe makes about 3 to 4 (½-pint) jars or glasses.

2 cans (6 oz. each) frozen orange juice concentrate, thawed
*1 package powdered fruit pectin**
2½ cups water
4½ cups sugar

In a large saucepan or preserving kettle, combine the thawed concentrate, water and pectin. Heat until bubbles begin to form around edge of pan. Stir in the sugar and heat to full rolling boil. Boil hard 1 minute, stirring constantly. Remove from the heat immediately and skim off the foam. Ladle into clean, hot ½-pint jars or sterilized jelly glasses to within ½ inch of each top. Wipe off the tops and threads of the jars, or tops of the glasses, with a damp cloth. Put prepared lids on the jars and seal as the manufacturer directs; seal glasses with paraffin. Process the jars in a boiling water bath for 5 minutes, or invert quickly so the hot jelly touches the lid, then stand them upright again.

**Follow the pectin package directions for exact boiling time. Some brands require 2 minutes.*

RHUBARB JAM PATTERSON

Candy orange slices and maraschino cherries spark this delightful rhubarb jam from Mrs. Vernon Patterson. Mrs. Patterson is a blue ribbon champion from Raymond, Washington. The recipe makes 10 to 12 (½-pint) jars.

2 quarts chopped rhubarb
8 cups sugar
1 jar (8 or 10 ounces) maraschino cherries, drained and chopped
20 candy orange slices, chopped (about 10 ounces or 2 cups)

Combine all the ingredients in a large preserving kettle. Boil them slowly about 30 minutes or until thick. Ladle the jam into clean hot ½-pint jars to within ½ inch of each jar's top. Wipe off the tops and threads of the jars with a damp cloth. Put on prepared lids and seal as the manufacturer directs. Process in a boiling water bath for 5 minutes.

HONEY SAUTERNE JELLY DAVIS

Allen C. Davis, Sr., of Smyrna, Georgia, makes this elegant jelly. When you taste it you'll understand why it won a prize at North Georgia State Fair. Try it with roast lamb, veal, pork or chicken, or as a glaze for ham. It is also a superb addition to afternoon teas. The recipe makes 4 to 5 (½-pint) jars or jelly glasses.

1½ cups sauterne
2 teaspoons grated orange peel
½ cup orange juice
2 tablespoons lemon juice
*½ package powdered fruit pectin (2 tablespoons) **
3 cups mild-flavored honey

Combine the sauterne, orange peel and juice, lemon juice and pectin in a large preserving kettle. Heat and stir over high heat to a full rolling boil; boil 5 minutes, stirring constantly. Remove from the heat and quickly skim off the foam with a slotted spoon. Quickly pour the jelly into clean, hot ½-pint jars or sterilized jelly glasses. Wipe off the tops and threads of the jars or glasses with a damp cloth. Seal the glasses with paraffin; seal the jars with two-piece, self-sealing lids as the manufacturer directs.

**Follow the pectin package directions for exact boiling time. Some brands require 2 minutes.*

PEACH JAM MOORE

This recipe comes from Donald E. Moore III, who had to interrupt his jam testing to become a member of the Syracuse University Ski Team. He is obviously a winner in jam competition as well as on the slopes. Mr. Moore says that Peach Jam was only his third attempt at making jam — but it won a blue ribbon at the New York State Fair. The recipe makes about 8 (½-pint) jars.

- 1 quart peeled, chopped or mashed ripe peaches
- 2¾ pounds sugar
- 1 jar (9 ounces) orange-flavored breakfast drink powder (Tang)
- ¼ cup lemon juice
- ½ bottle liquid fruit pectin

Combine the peaches, sugar, orange-flavored drink powder and lemon juice in a large preserving kettle. Mix well and heat to a full rolling boil. Boil hard 1 minute, stirring constantly. Remove the kettle from the heat and immediately stir in the liquid fruit pectin. Stir and skim for 5 minutes to prevent floating fruit. Ladle the jam into clean, hot ½ -pint jars to within ½ inch of each jar's top. Wipe off the tops and threads of the jars with a damp cloth. Put on prepared lids and seal as the manufacturer directs. Process in a boiling water bath for 5 minutes.

BERRY JAM VERCAUTEREN

Marie Vercauteren, a champion at the Colorado State Fair from Pueblo, Colorado, uses frozen berries for her jam. You can make up a couple of jars whenever the fancy strikes you. The recipe makes about 2 (1 pint) jars.

- 1 bag (1 pound 4 ounces) frozen unsweetened blackberries, strawberries, blueberries or raspberries
- 2½ cups sugar
- 2 tablespoons lemon juice

Combine the berries, sugar and lemon juice in a large preserving kettle. Heat and stir over medium-high heat until the sugar dissolves. Boil rapidly, stirring frequently to prevent sticking, until jam is thick, reaches 8°F above boiling, or drips in sheets from a spoon. Skim off the foam and ladle into clean, hot pint jars to within ½ inch of each jar's top. Wipe off the tops and threads of the jars with a damp cloth. Put on prepared lids and seal as the manufacturer directs. Process in a boiling water bath for 5 minutes.

RED CURRANT JELLY THOMPSON

This classic jelly can go onto breads and into jelly rolls, as well as into many meat and fruit sauces. The recipe won first place at the Ohio State Fair for Mrs. Billy Thompson, Lewis Center, Ohio. The recipe makes about 9 (½-pint) jars or jelly glasses.

3½ quarts red currants
1½ cups water
*1 box powdered fruit pectin **
7 cups sugar

Wash the currants and drain well. Remove the leaves but not the stems. Put the currants in a large preserving kettle and crush them with a potato masher or back of a large wooden spoon. Add the water; heat to boiling, then reduce the heat, cover and simmer 10 minutes. Pour the currant mixture into a jelly bag and let it hang until all the juice has dripped out (several hours or overnight). If you squeeze the bag to hurry the juice along, you must strain the juice through 4 layers of cheesecloth so it will not be cloudy. Measure 6½ cups juice into a large preserving kettle, adding water if necessary. Stir in the pectin. Heat and stir over high heat to a full rolling boil. Stir in the sugar and heat again to a full rolling boil; boil 1 minute, stirring constantly. Remove the kettle from the heat and quickly skim off the foam with a slotted spoon. Immediately pour the jelly into clean, hot ½-pint jars or sterilized jelly glasses. Wipe off the tops and threads of the jars or glasses with a damp cloth. Put on prepared lids and seal as manufacturer directs. Pour on paraffin to seal jelly glasses.

**Follow the pectin package directions for exact boiling time. Some brands require 2 minutes.*

RED PLUM JELLY WILSON

Ruth J. Wilson from Benton, Illinois shares this tasty, pretty jelly with us. It won a blue ribbon at the Illinois State Fair. The recipe makes about 4 (½-pint) jars.

4 to 4½ pounds firm but ripe red plums
1 to 1½ cups water
4 cups sugar

Wash the plums and put them in a preserving kettle with water. Heat to boiling, then simmer, covered, about 20 minutes or until the plums are very tender. Pour the plums into a strainer, reserving juice. Pour the juice into a jelly bag or 4 layers of cheesecloth and strain again. Measure 4 cups juice into a preserving kettle and heat to boiling, skimming off any foam that may form. When the juice is boiling add the sugar and stir until dissolved. Boil slowly, without stirring, about 15 minutes, then begin to test for the proper "jell." (The jelly tests done when it reaches 8°F above boiling or drips in a sheet from a spoon.) When done, immediately remove it from the heat. Skim off the foam. Ladle into sterilized jelly glasses to within ½ inch of top, or ladle into clean, hot jars to within ⅛ inch of each top. Wipe off the tops and threads of the jars and the inside rims of the glasses with a damp cloth. Put prepared lids on the jars and seal as the manufacturer directs. Invert jars quickly so hot jelly touches lid, then stand upright again. Seal the glasses with paraffin.

RIPE PAPAYA JAM PETERSON

This winning recipe comes from Floreta Peterson in the Virgin Islands, where papayas are plentiful. The recipe makes 8 (½ pint) jars.

6 to 8 medium-to-large ripe papayas
5 cups sugar
½ *cup lemon or lime juice*

Peel papayas and remove the seeds. Press the fruit through a coarse sieve, or grind, using a coarse blade. Measure 6 cups pulp into a large saucepan or preserving kettle. Heat to boiling, then boil, stirring constantly, until thick. Stir in the sugar and lemon juice and boil until thick and clear, stirring constantly. Ladle into clean, hot, ½-pint jars to within ½ inch of each jar's top. Wipe off the tops and threads of the jars with a damp cloth. Put on prepared lids and seal as the manufacturer directs. Process in a boiling water bath for 5 minutes.

HEAVENLY JAM STONEHOUSE

Strawberry, pineapple and orange combine in Dorothy Stonehouse's other-worldly winner from Presque Isle, Maine. The recipe makes 6 to 8 (½-pint) jars.

1 quart strawberries
1 can (1 pound 4½ ounces) crushed pineapple, drained
2 to 3 tablespoons grated orange peel
2 tablespoons lemon juice
*1 package powdered fruit pectin **
5 cups sugar

Hull the berries and crush them in a large preserving kettle. Add the pineapple, orange peel, lemon juice and pectin. Heat to a full rolling boil. Stir in the sugar and heat again to a full rolling boil. Boil hard 1 minute, stirring constantly. Ladle the jam into clean hot ½-pint jars to within ½ inch of each jar's top. Wipe off the tops and threads of the jars with a clean damp cloth. Put on the prepared lids and seal as the manufacturer directs. Process in a boiling water bath for 5 minutes.

**Follow the pectin package directions for exact boiling time. Some brands require 2 minutes.*

DEWBERRY JAM BIGLER

If you can locate some dewberries be sure to try Mrs. H.J. Bigler's blue ribbon winner from the Utah State Fair. The recipe makes about 4 (½-pint) jars.

1½ quarts dewberries
*½ package powdered fruit pectin**
4 cups sugar

Wash the berries. Grind or mash them to a pulp. Measure 3 cups mashed berries into a large saucepan or preserving kettle. Add the pectin. Heat to a full rolling boil, then stir in the sugar and heat again to a full rolling boil. Boil hard for 1 minute. Remove from the heat immediately. Stir and skim for 5 minutes to prevent floating fruit. Ladle into clean, hot ½ pint jars to within ½ inch of each jar's top. Wipe off the tops and threads of jars with a damp cloth. Put on prepared lids and seal as the manufacturer directs. Process in a boiling water bath for 5 minutes.

**Follow the pectin package directions for exact boiling time. Some brands require 2 minutes.*

RHUBARB JELLY HAYNES

If there is no rhubarb in your garden, by all means go out and buy some so you can make this beautiful pink jelly. It is from Myrna Haynes, a blue ribbon winner at the Colorado State Fair. The recipe makes 8-9 (½ pint) jars or glasses.

3½ pounds trimmed rhubarb
2½ cups water
*1 box powdered fruit pectin**
7 cups sugar

Cut the rhubarb into small pieces and put them in a large kettle along with the water. Heat to boiling, then boil gently until the rhubarb is soft but not mushy. Pour the rhubarb into a jelly bag or 4 layers of cheesecloth in a colander. Hang the bag, or tie ends of the cheesecloth together to form bag and hang. Let the juice drip into bowl overnight. Measure 5 cups juice into a large preserving kettle. (If you do not have quite enough juice, pour boiling water over the rhubarb that remains in jelly bag.) Stir the pectin into the juice and heat to a full rolling boil. At once, add the sugar and heat again to a full rolling boil, stirring constantly. Boil hard 1 minute. Immediately remove the jelly from the heat. Skim off the foam. Ladle it into sterilized jelly glasses to within ½ inch of each top, or ladle into clean, hot ½-pint jars to within ⅛ inch of each top. Wipe off the tops and threads of the jars and the inside rims of the glasses with a damp cloth. Put prepared lids on the jars and seal as the manufacturer directs. Invert the jars quickly so the hot jelly touches the lid, then stand them upright again.

**Follow the pectin package directions for the exact boiling time. Some brands require 2 minutes.*

GINGER PEARS CASIDA

Almost a marmalade, but spiked with ginger, this winning recipe from Mrs. Orin Casida of Pueblo, Colorado, took a blue ribbon at the Colorado State Fair. The recipe makes about 5 (1-pint) jars.

5 pounds hard pears
3 cups water
3 lemons
Boiling water
10 cups sugar
½ cup (about 4 ounces) preserved ginger, minced

Pare and core the pears and cut them in thin slices. Cook the pears in a large preserving kettle with the 3 cups water just until tender, about 5 to 10 minutes. Meanwhile, cut the lemons in quarters and then in thin slices. Pour boiling water over the lemons to cover: let them stand 5 minutes and then drain. Add the lemon to the pears along with the sugar and ginger. Simmer until the preserves are thick, reach 8°F above boiling or drip in a sheet from a spoon. Stir and skim for 5 minutes to prevent floating fruit. Ladle the preserves into clean, hot pint jars to within ½ inch of each jar's top. Wipe off the tops and threads of the jars with a damp cloth. Put on prepared lids and seal as the manufacturer directs. Process in a boiling water bath for 5 minutes.

PEACH JAM MACDOUGAL

Oranges and maraschino cherries, as well as peaches, make up this blue ribbon winner from the Northern Maine Fair held at Presque Isle. Mrs. Ronald MacDougal sent it to us and we concur with the Maine judges — it's great! The recipe makes about 8 pints, 16 (½-pint) jars.

4 to 5 pounds peaches
5 oranges
10 cups sugar
1 cup chopped maraschino cherries

Wash the peaches. Plunge or dip them into boiling water for 30 to 60 seconds, then dip them in cold water. Peel, pit and slice or cut the peaches in chunks. Grind the chunks with the coarse blade of a food grinder. Remove the thin outer portion of peel from the oranges and reserve. Cut off the white portion of the peel and discard. Slice the oranges into chunks and grind along with the reserved out portion of the peel. Mix the ground oranges with the peaches and measure (you should have about 10 cups).

Combine the fruit mixture with the sugar in a large preserving kettle. (Add another cup or two of sugar if you prefer a little sweeter jam.) Cook and stir over medium high heat until the mixture is 8°F above boiling or drips in a sheet from a spoon. Stir and skim for 5 minutes to prevent floating fruit. Stir in the maraschino cherries. Ladle the jam into clean, hot ½-pint or pint jars to within ½ inch of each jar's top. Wipe off the tops and threads of the jars. Put on prepared lids and seal as the manufacturer directs. Process in a boiling water bath for 5 minutes.

PEACH JAM SPRAGUE

Another favorite recipe from Aroostook County, Maine, is given a special touch by Alice Sprague, who adds ground oranges and maraschino cherries to her peach jam. Try some on your own homemade bread. The recipe makes 5 (1-pint) jars.

5 pounds ripe peaches
5 oranges
Sugar
1 cup chopped maraschino cherries

Peel and pit the peaches and cut them into chunks. Cut the oranges into chunks and remove the seeds. Grind the peaches and oranges together using the coarse blade of a food grinder. Measure the ground fruit into a large preserving kettle. Add 1½ cups sugar for each cup fruit. Heat to boiling, then boil slowly, stirring frequently to prevent sticking, until the mixture thickens, reaches 8°F above boiling or drips in a sheet from a spoon. Stir in the cherries. Ladle the jam into clean, hot pint jars to within ½ inch of each jar's top. Wipe off the tops and threads of the jars with a damp cloth. Put on prepared lids and seal as the manufacturer directs. Process in a boiling water bath for 5 minutes.

PARADISE PEAR JAM BESICH

Mary J. Besich's divine jam won a prize at a Montana State Fair. Paradise Pear Jam makes a great holiday gift or hostess gift. Try it on hot muffins, cornbread or popovers. The recipe makes about 5 (½-pint) jars.

- *2 pounds ripe but firm pears*
- *1 can (8¼ ounces) crushed pineapple*
- *1 orange*
- *1 lemon*
- *½ cup finely chopped citron*
- *¼ chopped maraschino cherries*
- *5 cups sugar*
- *1 box powdered fruit pectin**

Wash, peel and core the pears. Chop them finely, or grind in food grinder or blender. Cut off the colored outer portion of the orange and lemon peel and chop finely. Cut away the white portion of peel and discard. Remove the seeds, then chop the orange and lemon and add them to the pears in a large saucepan or preserving kettle, along with all the other fruits. Stir in the pectin. Heat and stir over high heat until the mixture comes to a full rolling boil. Stir in the sugar and heat again to a full rolling boil; boil 1 minute, stirring constantly. Remove the kettle from the heat and stir, skimming with a slotted spoon for 5 minutes to keep the fruit from floating. Ladle into clean hot ½-pint jars to within ½ inch of each jar's top. Wipe off the tops and threads of jars with a damp cloth. Put on prepared lids and seal as the manufacturer directs. Process in boiling water bath for 5 minutes.

**Follow the pectin package directions for exact boiling time. Some brands require 2 minutes.*

PINEAPPLE STRAWBERRY JAM BANKER

Doreen Banker of Pawcatuck, Connecticut, uses either fresh or canned pineapple and fresh or frozen strawberries to make this winning recipe a favorite for any season. This recipe won her a trip to the National 4-H Congress and a State Honor Certificate. The recipe makes about 5 (½-pint) jars.

1 ripe fresh pineapple or 1 can (1 pound 13 ounces) crushed pineapple, drained
1 quart fresh strawberries or 1 bag (1 pound 4 ounces) frozen strawberries
7 cups sugar
½ bottle liquid fruit pectin

Pare and core the pineapple, cut out the eyes. Grind or chop it finely. Hull and crush the berries. You should have 1 quart of prepared fruit. In a large preserving kettle, combine the fruit and sugar. Heat them to a full rolling boil. Boil hard for 1 minute, stirring constantly. Remove the kettle from the heat; stir in the liquid fruit pectin. Stir and skim for 5 minutes to prevent floating fruit. Ladle the jam into clean, hot ½-pint jars to within ½ inch of each jar's top. Wipe off the tops and threads of the jars with a damp cloth. Put on the prepared lids and seal as the manufacturer directs. Process in a boiling water bath for 5 minutes.

RED OR BLACK RASPBERRY PRESERVES SANDERS

These preserves are wonderful on hot homemade bread or muffins, rolled up inside a jelly roll, atop pancakes or spooned over a fresh fruit cup! These preserves won top prize at the Macoupin County Fair for Mrs. George Sanders of Medora, Illinois. The recipe makes 6 to 7 (½-pint) jars.

5 pounds raspberries (3 heaping quarts, if you pick them yourself)
6 cups sugar

Gently wash the berries and pick over to remove any bad ones. Heat the berries slowly in a large preserving kettle until the juice begins to flow. Stir in the sugar. Boil rapidly, stirring occasionally to prevent sticking, for 20 minutes. Ladle the preserves into clean, hot jars to within ½ inch of each jar's top. Wipe off the tops and threads of the jars with a damp cloth. Put on the prepared lids and seal as the manufacturer directs. Process in a boiling water bath 5 minutes.

NOTE: For Black Raspberries, Mrs. Sanders suggests 8 cups of sugar.

STRAWBERRY JAM SCHAACK

This toothsome, quick-cooking jam recipe calls for frozen berries. Susan Schaack took best in class with it at the Passaic County 4-H fair in New Jersey. The recipe makes about 6 (½-pint) jars.

1 bag (1 pound 4 ounces) frozen unsweetened strawberries
*½ package powdered fruit pectin (2 tablespoons)**
2 tablespoons water
2 tablespoons lemon juice
3 cups sugar

Thaw and crush the fruit in a large preserving kettle. Add the pectin, water and lemon juice. Heat to a full rolling boil. Stir in the sugar and heat again to a full rolling boil. Boil hard 1 minute, stirring constantly. Remove the kettle from the heat and stir and skim for 5 minutes to prevent floating fruit. Ladle the jam into clean, hot (½-pint) jars to within ½ inch of each jar's top. Wipe off the tops and threads of the jars with a clean damp cloth. Put on prepared lids and seal as the manufacturer directs. Process in a boiling water bath for 5 minutes.

**Follow the pectin package directions for exact boiling time. Some brands require 2 minutes.*

FIG PRESERVES McFARLAND

Linda McFarland, from Ocilla, Georgia, was a winner at the Irwin County Fair with this recipe. Lucky you, if you have a fig tree or a friend with one. The recipe makes about 12 (1-pint) jars.

6 quarts figs
1 cup baking soda
6 quarts boiling water
8 cups sugar
2 quarts water

Wash and pick over the figs and discard any that are overripe or broken. Sprinkle the baking soda over the figs, then pour the boiling water over them. Let them stand 15 minutes. Drain well, then rinse thoroughly in clear, cold water. Let them drain while preparing the syrup. Combine the sugar and the 2 quarts water in a large preserving kettle and heat to boiling. Boil 5 minutes; skim off any foam. Gradually add the drained figs to the hot syrup, heat through, then boil until the figs are clear, about 2 hours. Carefully lift the figs out of the syrup with a slotted spoon and place them in a shallow pan. If the syrup is still thin, boil for several minutes until it is thick and syrupy. Pour the syrup over the figs, being sure to coat each one. Let them stand overnight. Pack into clean, hot jars with the stems up. Pour in the syrup to within ½ inch of each jar's top. Wipe off the tops and threads of the jars. Put on prepared lids and seal as the manufacturer directs. Process in a boiling water bath for 15 minutes.

STRAWBERRY PRESERVES KUNKEL

A family favorite, this recipe won first-prize at the Du Quoin State Fair for Mildred Kunkel of Tamaroa, Illinois. Plan to put up plenty of these preserves — you'll want them on your breakfast table all-year long. The recipe makes 9 (½-pint) jars.

3 quarts strawberries
½ cup drained crushed pineapple
*1 package powdered fruit pectin**
7 cups sugar

Wash and hull the strawberries. Slice or halve them into a large preserving kettle. Add the pineapple and pectin. Heat and stir over high heat to a full rolling boil. Stir in the sugar and heat again to a full rolling boil; boil 2 minutes. Remove the preserves from the heat and stir and skim for 5 minutes to keep the fruit from floating. Ladle into clean, hot ½-pint jars to within ½ inch of each jar's top. Wipe off the tops and threads of the jars with a damp cloth. Put on prepared lids and seal as the manufacturer directs. Process in a boiling water bath for 5 minutes.

**Follow the pectin package directions for exact boiling time. Some brands require 2 minutes.*

OLD FASHIONED TOMATO PRESERVES BESICH

These preserves won raves for Mary J. Besich of Black Eagle, Montana. They can win praise for you, too. They are delicious on corn muffins or hot, homemade whole wheat bread. The recipe makes 4 (½-pint) jars.

About 2 pounds ripe tomatoes
*1 packaged powdered fruit pectin**
2 teaspoons grated lemon peel
¼ cup lemon juice
¼ teaspoon ground cloves
¼ teaspoon ground allspice
3 cups sugar

Wash and peel the tomatoes. Chop them coarsely, measure 1 quart and put it into a large saucepan. Heat it to boiling, then reduce the heat and simmer 10 minutes, stirring occasionally to prevent sticking. Stir in the pectin, lemon peel, lemon juice and spices. Heat to a full rolling boil. Stir in the sugar and heat again to a full rolling boil; boil 1 minute, stirring constantly. Remove the saucepan from the heat and stir and skim for 5 minutes to keep the fruit from floating. Ladle the preserves into clean, hot ½-pint jars to within ½ inch of each jar's top. Wipe off the tops and threads of the jars with a damp cloth. Put on prepared lids and seal as the manufacturer directs. Process in a boiling water bath for 5 minutes.

NOTE: In place of the ground spices, you may use 12 whole cloves and 12 whole allspice tied in a cheesecloth bag.

**Follow the pectin package directions for exact boiling time. Some brands require 2 minutes.*

STRAWBERRY FIG PRESERVES CHILDS

Vicki Childs, a Grady County, Georgia, 4-H'er, says "If you didn't know the difference you'd believe you were eating strawberries instead of figs." This recipe has won her blue ribbons in six different fairs. The recipe makes about 8 (½-pint) jars.

6 cups ripe, mashed figs
6 cups sugar
4 packages (3 oz. each) strawberry-flavored gelatin

Heat the figs and sugar to boiling in a large preserving kettle. Stir in the gelatin and boil slowly for 20 to 30 minutes or until thickened. Ladle into clean, hot ½-pint jars to within ½ inch of each jar's top. Wipe off the tops and threads of the jars with a damp cloth. Put on prepared lids and seal as the manufacturer directs. Process in a boiling water bath for 5 minutes.

UNCOOKED STRAWBERRY PRESERVES LUEHRMAN

These preserves have extra fresh flavor, since they are not cooked. But, you must keep them in the refrigerator. Lois Luehrman shares this winning recipe. She represented Missouri at the National 4-H Congress in Food Preservation. Miss Luehrman says this recipe has become her family's favorite. She serves it on fresh yeast rolls. No wonder people keep asking her for the recipe. The recipe makes about 3½ cups preserves. You can store it in small plastic containers with lids, baby food jars or refrigerator containers — anything with a tight-fitting lid that can go into the refrigerator or freezer.

3 packages (10 ounces each) frozen sliced strawberries, thawed
3 cups sugar
4 teaspoons lemon juice
¾ cup water
*1 package powdered fruit pectin**

In a large bowl, combine the strawberries, sugar and lemon juice. In small saucepan, combine the pectin and water and heat to boiling; boil 1 minute. Pour the hot pectin over the berry mixture and stir for 2 to 3 minutes. Pour the preserves into a clean container that has a tight-fitting lid; cover and let the preserves stand until set, up to 24 hours. Store them in the refrigerator or freezer.

**Follow the pectin package directions for exact boiling time. Some brands require 2 minutes.*

MARY'S FAVORITE CONSERVE

Because you can use canned fruit or fresh, you can make this thick, sweet treat any time of year. Mrs. Charles Hildreth of Fairborn, Ohio, won a prize at the Ohio State Fair for her recipe. The recipe makes about 7 (½-pint) jars.

2 cups diced canned or fresh pears
2 cups diced canned or fresh peaches
1 cup drained pineapple chunks
4 to 5 cups sugar (less for canned fruit)
3 to 4 tablespoons lemon juice
1 medium orange, peeled and chopped fine
½ cup sliced maraschino cherries
½ cup coarsely chopped walnuts

In a large preserving kettle, combine all the ingredients except the cherries and nuts. Boil until the desired thickness, about 1 hour, stirring frequently. Stir in the cherries and nuts. Ladle the conserve into clean, hot jars to within ½ inch of each jar's top. Wipe off tops and threads of the jars with a damp cloth. Put on prepared lids and seal as the manufacturer directs. Process in boiling water bath for 5 minutes.

ORANGE MARMALADE LEEK

Sheila Leek from Blooming Prairie, Minnesota, says this is one of her favorite recipes. Her jam and jelly recipes helped her win the Minnesota 4-H Food Preservation Project. The recipe makes about 8 (½-pint) jars.

4 oranges
2 lemons
2½ cups water
⅛ teaspoon baking soda
*1 box powdered fruit pectin **
6½ cups sugar

Cut the peel from the oranges and lemons. Cut off half the inner white portion of the peel and discard. Cut the remaining peel into fine slices. Combine the sliced peel in a saucepan with the water and soda and simmer for 20 minutes. Cut the oranges and lemon pulp into fine chunks or chop them in a blender. Add the chopped oranges and lemons to the peel in the pan and simmer for 10 minutes. Measure 4 cups fruit mixture and return them to the saucepan or a large preserving kettle. Add the pectin and heat to a full rolling boil. At once, add the sugar and heat to a full rolling boil, stirring constantly. Boil hard 1 minute. Immediately remove the pan from the heat. Skim off the foam. Stir and skim for 5 minutes to prevent floating fruit. Ladle into clean, hot ½-pint jars to within ½ inch of each top. Wipe off the tops and threads of the jars with a damp cloth. Put on prepared lids and seal as the manufacturer directs. Process in a boiling water bath for 5 minutes.

**Follow the pectin package directions for exact boiling time. Some brands require 2 minutes.*

GRAPE BUTTER MEYERS

Thick, rich and sweet, this winner from the Macoupin County Fair in Illinois comes from Leta M. Meyers. Grape butter is a great way to preserve the fruits of your Concord grape vine. The recipe makes about 9 to 10 (½-pint) jars.

About 4 to 5 pounds Concord grapes
½ cup water
*1 box powdered fruit pectin**
7½ cups sugar

Crush the grapes in a large preserving kettle. Add water and simmer, covered, for 10 minutes. Press the grapes through a food mill, sieve or colander to remove the seeds. Measure the pulp into a large preserving kettle. Stir in the pectin and heat to a full rolling boil. At once, add the sugar and heat to a full rolling boil, stirring constantly. Boil hard for 1 minute. Immediately remove the kettle from the heat. Skim off the foam. Ladle the butter into clean, hot jars to within ½ inch of each top. Wipe off the tops and threads of the jars with a damp cloth. Put on prepared lids and seal them as manufacturer directs. Process in a boiling water bath for 5 minutes.

**Follow the pectin package directions for the exact boiling time. Some brands require 2 minutes.*

PEACH MARMALADE BIGLER

The blue ribbon Mrs. H. J. Bigler took for this orange-peach delight is just one of 90 Utah State Fair ribbons she has won in only three years. The recipe makes about 3 (1-pint) jars or 6 (½-pint) jars.

10 large ripe peaches
2 large oranges
5 cups sugar
2 tablespoons lemon juice

Peel, pit and chop the peaches into a large preserving kettle. (Grate the peel of one orange, if desired, and add it to the peaches.) Peel the oranges and chop, removing the tough membrane, if necessary. Stir in the sugar and lemon juice and cook and stir over medium-high heat about 30 to 45 minutes, or until the mixture is thickened, is 8°F above boiling or drips in a sheet from a spoon. Stir and skim for 5 minutes to prevent floating fruit. Ladle the marmalade into clean, hot ½-pint or pint jars to within ½ inch of each jar's top. Wipe off the tops and threads of the jars with a damp cloth. Put on prepared lids and seal as the manufacturer directs. Process in a boiling water bath for 5 minutes.

PEACH BUTTER WELLER

Mrs. Roy Weller of Carlinville, Illinois, won a blue ribbon for her peach butter at the Macoupin County Fair. Her butter recipe is for just a few peaches, or as many batches as you can handle. Remember that 1 pound of fresh peaches will yield about 2½ cups peeled and sliced or about 2 half-pint jars of butter. Peaches that aren't quite perfect enough for canning as halves or slices are good candidates for butter.

Peaches
Sugar (See preparation below for amount)

Wash the peaches. Plunge or dip them into boiling water for

30 to 60 seconds, then dip them in cold water. Peel, pit and slice or cut the peaches in chunks into a large preserving kettle. Add ½-cup water or just enough to barely cover the bottom of the kettle. Cook over medium-high heat until the peaches are soft, adding another tablespoon or two of water, if necessary to prevent sticking.

Remove the kettle from the heat and press the peaches through a sieve, food mill or blend in a blender until smooth. Measure the puree and return it to the kettle. Add ½-cup sugar for each cup puree. (If your tooth is a little sweeter, make it ⅔ cup sugar for each cup puree.) Cook over medium heat, stirring often to prevent sticking, until the butter is very thick. Ladle into clean, hot ½-pint jars to within ½ inch of each jar's top. Wipe off the tops and threads of the jars with a damp cloth. Put on prepared lids and seal as the manufacturer directs. Process in boiling water bath for 10 minutes.

Drying

Drying food was a necessity for the early Americans. Fresh produce did not miraculously appear at a corner store. No fast food chains stretched from sea to shining sea to sustain the traveler. Vegetables, fruits and herbs were regularly dried by farm wives until canning and freezing became common. Grapes were dried in bunches hung from barrel hoops. Herbs were bunched and hung upside down from the beams. In the Southwest, colorful strings of dried peppers provided the major seasoning. And, everywhere dried corn was the staple for hundreds of dishes — including corn mush, ashe cake, jonnycake (journey cake), hasty pudding (corn mush with maple syrup) and, in the Southwest, the tortilla.

The only people who seemed to have spurned the Indian's saving gift of corn were the brides imported from France in 1727 for the French settlers in New Orleans. They refused to cook corn — in fact, they rebelled and told the governor that they would only cook with fine French wheat flour. Possibly the first cooking school in America was organized to help those homesick French women overcome their distaste for corn.

To give you an idea of the trouble it was to turn corn into an all-purpose edible, we offer a recipe for drying your own corn (white, yellow or blue). If you want to re-enact the authentic early American routine, forego the oven and electric grinder.

Use grindstones, or — even more difficult — locate a gristmill.

Dried corn and dried snap beans or lima beans can be mixed together to make another famous American Indian dish — succotash. Succotash was a light weight, high-protein food that could be packed along for extensive journeys. Dried foods have come into vogue again as portable foods for back-packing and camping. Commercially-dried and packaged foods for back-packing are expensive; you can save a considerable amount if you dry and package your own fruits and vegetables and make your own beef jerky (see Smoking). Dried fruits and nuts also make good energy-filled snacks for long hikes or bike rides.

Modern Drying Methods

Drying preserves food by removing its moisture, thus cutting off the water supply for spoilage causing bacteria. Drying is probably the first method of food preservation used by man. The sun, of course, is the dryer our ancestors used. If you live where the sun shines long — California or Arizona — you too can dry fruits and vegetables in the sun. But, for those in the remaining states, drying is a little easier with an oven or simple-to-make dryer.

Oven drying is more expensive than sun drying, but may be more reliable, depending upon where you live. Drying does take attention. You will need to be at home while you are drying food, and it can take many hours. Food must be stirred, tray positions must be changed and temperatures must be checked.

Pre-treatment

Fruits need a pre-treatment before drying to prevent them from darkening. The recipes tell you exactly how to pre-treat. Vegetables and some fruits need blanching in steam or boiling water to stop enzyme action. The blanching is extremely brief. Blanched vegetables to be dried do not need to be chilled, as do blanched vegetables for the freezer.

Temperature

Drying in an oven must be done at low, even temperatures, usually around 120°F. This provides just enough heat to dry the food without cooking it. Maintaining this temperature evenly, with air circulation, is the trick to successful drying

You will need to get an oven thermometer that registers as low as 100°F and then you will need to do some experimenting with your oven, its control, door, racks, and perhaps with a fan.

Electric ovens sometimes have a "Low" setting or "Warm" setting that may be the right temperature for drying. Or, sometimes the light bulb in the oven creates enough heat. Open the oven door slightly, or alot, depending on how much outside air you need to get the temperature right. It may help to set a fan on "Low" and aim its breeze across the front of the open or ajar oven door. You will have to test and try, and test and try again, to see just which arrangement works best.

The pilot light of a gas oven may provide just the right amount of heat, or you may be able to set the heat control at 120°F. Again, it is a matter of trial, test and judgement on your part. Put in the oven thermometer, wait about 10 minutes and take a reading. If it is not right, make adjustments in door opening or temperature control and try again. A fan may not work well for a gas oven since it can blow out the pilot light.

A simple-to-make drying box (directions in this section) is another possibility. You may find other alternatives. Perhaps you have radiators that, in the winter, send out enough heat to dry fruits and vegetables. Or, perhaps in the summer your attic is hot and dry enough.

How to Make a Box Dryer

A hardware or discount store should have everything you need to make this simple dryer.

1. **A metal cookie sheet** with sides or a jelly roll pan is needed to hold the food.
2. **An empty cardboard box** that has the same top dimensions as the cookie sheet forms the drying box. The sheet should just fit on top of the box, or the rims of the sides should rest on the edges of the open-topped box.
3. **A box of heavy-duty or extra-wide aluminum foil** is used to line the box.
4. **A small can of black paint** is used to paint the bottom of the cookie sheet; buy a spray can or get a small brush.
5. **A 60 watt light bulb and socket** attached to a cord and plug provide the heat.

Line the inside of the box with foil, shiny side up. Cut a tiny notch in one corner for the cord to run out. Set the light fixture in the center, resting it on a crumpled piece of foil. Paint the bottom of the cookie sheet black and let it dry.

Prepare the food as the recipe directs. Spread it in an even, single layer on the black-bottomed cookie sheet. Then put the sheet in place on top of the box. Plug in the light bulb to preheat the box and dry until the food tests done. (Each recipe specifies how the food should feel when it is dry.) If you are making more than one sheet of food, obviously you will have to make more than one drying box. Do not prepare more food than you can dry at one time.

Basic Equipment

1. **An oven thermometer** that will read as low as 100°F is a necessity for oven drying.
2. **Sharp, stainless steel knives,** that will not discolor fruits or vegetables, will be needed for thin-slicing, peeling and halving.
3. **A cutting board** is essential for chopping and slicing. Be sure to scrub the board thoroughly before and after using.
4. **Measuring cups and spoons** are needed for preparing ascorbic acid.
5. **Baking or cookie sheets** hold food in the oven or the box dryer. Sheets without any edges are the best: they allow hot air to circulate around to all sides of the food.
6. **A blancher** is needed for pre-treatment of all vegetables and some fruit. Use a ready-made blancher, or make one from a deep pot with a cover and colander or basket that will fit down inside the pot. For steam blanching you will need a rack or steamer basket.
7. **A long, flexible spatula** helps you stir the food on the cookie sheet so it will dry more evenly.
8. **Storage containers** must be air tight. Use plastic, glass or metal containers that have tight-fitting lids — coffee cans lined with plastic bags, jars, freezer containers, refrigerator-ware. Double plastic bags closed tightly with string, rubber bands or twist ties may also be used.

Basic Ingredients

Each recipe tells you what to look for when picking or shopping. Your goal should be to find perfect fruits and vegetables. Fruits must be fully ripe, but not soft. Vegetables should be tender and mature, but not woody. All must be very fresh. Every minute from harvest to the drying tray can mean loss of flavor and quality, so hurry. Do not use any produce with bad spots. Remember that one bad apple, indeed, can spoil the whole barrel — or drying tray.

Blanching

For boiling water blanching: Heat 1 gallon water to boiling in a blancher. Put no more than 1 pound or 4 cups prepared vegetable or fruit into the blancher's insert, colander or strainer and carefully lower it into the boiling water for time given in the recipe.

For steam blanching: Pour enough water in the blancher to cover the bottom, but not touch the insert. Heat to boiling. Arrange the prepared vegetables in a single layer in the blancher's insert; put them in the blancher over boiling water, cover tightly and steam for the time given in the recipe. You can use any large pot or kettle for steam blanching by putting a rack about three inches above the bottom to hold the vegetables in the steam and up out of the boiling water. You can put the vegetables in a cheesecloth bag to keep the pieces together during blanching, if you wish.

When Is It Dry?

In most other forms of food preservation, time is your guide to when the food is done. This is not so with drying. Time can vary considerably — from less than 12 to more than 24 hours — depending on the food, how thin it is sliced, the heat, humidity and drying temperatures. There is no way to give an accurate time guide. Instead, you must feel the pieces of fruit or vegetable and judge their dryness by texture and consistency. Each recipe gives you a description of the "dry test" for each food.

Conditioning

Foods do not always dry evenly, nor do they dry at exactly the same rate for each piece or slice. To even up the drying process, put the cooled, dried food into a large deep container (a crock or clean dishpan). Keep it in a warm dry room, lightly covered with cheesecloth to keep out insects, and stir once or twice a day for several days (up to 10 days). After conditioning, give the fruit or vegetables one final treatment to get rid of any insects or insect eggs. Either put the dried food in the freezer for a few hours, or heat it on a cookie sheet in the oven at 150°F to 175° for 10 to 20 minutes.

Storage

Keeping out air and moisture is the secret to good dried foods. Store dried fruits and vegetables in a cool, dark, dry place. You may store dried fruits and vegetables in the refrigerator or freezer, if you wish, but why take up that valuable space with something that will keep at cool, room temperatures?

Rehydration

Dried fruits are often used as is, without returning to them the water they have lost. Some dried fruits, and all dried vegetables except onions, must be refreshed (soaked in water) before using. The soaking time depends on the food; soaking times range from 30 minutes to an hour or two. Dried foods will absorb about 1½ to 2 times their volume in water. Always cook dried fruits and vegetables in the water they have soaked in. That water is full of nutrients and should be used, never thrown away.

Basic Steps for Fruit

1. Set out all the ingredients and equipment. Wash and dry all equipment, countertops, working surfaces and your hands.
2. Select fruits that are ripe, tender and fresh.
3. Preheat the oven or box dryer so it will be around 120°F when the fruit is ready.
4. Wash the fruit well, scrubbing firm fruits with a brush. Handle them gently to avoid bruising.
5. Prepare an ascorbic acid mixture, if directed in the recipe, mixing crystalline or powdered ascorbic acid with cold water (the recipes will give you exact amounts)! Start the blancher heating, if blanching is needed. Fill the blancher ½ full of water, then start heating.
6. Prepare the fruit as the recipe directs, sprinkling it with ascorbic acid solution (if needed) as it is cut and peeled. Blanch, if the fruit requires it.

7. Arrange cut fruit in a single, even layer on a cookie sheet or sheets. Do not crowd the fruit on the sheet and do not prepare more than you can accommodate in the oven or dryer at one time.
8. For oven drying, put an oven thermometer toward the back of the tray. Put the tray on the top shelf in a preheated oven and maintain an oven temperature at about 120°F.
9. For box drying, turn on the light bulb to preheat the box. Place the tray on top of the box.
10. For both oven and box drying, check the tray often and stir the fruit on the tray, moving outside pieces to center. For oven drying, turn the tray from the front to the back and, if making more than one tray, change the trays from shelf to shelf for even drying. Check the trays more frequently during the last few hours of drying to prevent scorching.
11. When completely dried, remove the tray of fruit from the oven or box and let it stand until completely cooled.
12. Turn the dried fruit into a deep container, cover lightly with cheesecloth and condition, stirring once or twice a day, for several days.
13. Pack into airtight containers or double plastic bags and store in a cool, dark, dry place for up to a year.
14. Use the fruit as is, or rehydrate by putting it in a bowl, pouring in just enough boiling water to cover and letting the fruit stand for several hours. Use soaking liquid as you would fruit juice.

Recipes

APPLES

1. Have on hand a measuring cup, measuring spoons, knife, cookie sheets and an oven thermometer (for oven drying).
2. Choose perfect, tart cooking apples. Wash them well.
3. Combine 1 cup cold water with 2½ teaspoons crystalline or powdered ascorbic acid.
4. Peel the apples, core, cut them crosswise into slices, or dice.

5. Sprinkle the apples with the ascorbic acid mixture as you peel and cut them.
6. Arrange the apple pieces in a single, even layer on cookie sheets.
7. Dry until leathery and pliable, usually around 12 hours. Smaller pieces will be crisp.

APRICOTS

1. Have ready a measuring cup, measuring spoons, knife, cookie sheets and an oven thermometer (for oven drying).
2. Choose ripe, perfect apricots. Wash them well and drain.
3. Combine 1 cup cold water with 1½ teaspoons crystalline or powdered ascorbic acid.
4. Halve the apricots and remove the pits. Sprinkle them with ascorbic acid mixture as you halve and pit them.
5. Arrange the apricot halves, cup side up in a single layer on cookie sheets.
6. Dry until leathery and pliable, usually around 12 hours.

FIGS

1. Have on hand a blancher, cookie sheets and an oven thermometer (for oven drying).
2. Choose fully tree-ripened figs. Wash them well and drain.
3. Blanch figs in boiling water 45 seconds to split the skins. Drain well.
4. Halve only if large; otherwise leave them whole.
5. Spread the figs in an even layer on the cookie sheets.
6. Dry until leathery, usually around 12 hours. The inside of the figs should still be slightly sticky.

GRAPES

1. Have ready a blancher, cookie sheets and an oven thermometer (for oven drying).

(Continued On Next Page)

2. Choose seedless Thompson grapes that are ripe and perfect. Leave them on their stalk.
3. Wash them well and drain.
4. Blanch in boiling water about 5 seconds, just to split the skins.
5. Drain well and remove the grapes from the stalk, but do not stem.
6. Arrange single layers of grapes on cookie sheets.
7. Dry until wrinkled, usually around 12 hours. Grapes will be raisins, of course.
8. Stem when dried.

PEACHES

1. Have ready a large pan, colander, knife, measuring cup, measuring spoons, cookie sheets and an oven thermometer (if not using a drying box).
2. Choose fully ripe but not soft, perfect peaches.
3. Put the peaches in the colander and then in boiling water just long enough to loosen their skins, then dip them in cold water.
4. Combine 1 cup cold water with 1½ teaspoons crystalline or powdered ascorbic acid.
5. Peel peaches, halve and pit. Leave them in halves or slice, sprinkling with the ascorbic acid mixture as you peel, halve and slice.
6. Arrange the peaches in a single layer (cup side up, if cut in halves) on cookie sheets.
7. Dry until leathery and pliable, usually around 12 hours.

PLUMS AND PRUNE-PLUMS

1. Have ready a blancher, knife, cookie sheets and an oven thermometer (for oven drying).
2. Choose ripe, perfect plums or prune-plums.
3. Wash them well and drain.
4. Blanch in boiling water for 45 seconds, just to split skins.
5. Drain well, then halve and pit, if desired, or leave whole.
6. Arrange in a single layer on cookie sheets.
7. Dry until firm, leathery and pliable — usually around 12 hours.

Basic Steps for Vegetables

1. Set out all ingredients and equipment. Wash and dry all equipment, countertops, working surfaces and your hands.
2. Select vegetables that are freshly picked, tender and just mature enough to eat.
3. Preheat the oven or dryer so it will be around 120°F when the vegetables are ready.
4. Wash the vegetables thoroughly, scrubbing with a brush if necessary but handling them gently to avoid bruising.
5. Put a blancher on the range. For steam blanching, fill with just enough water to cover the bottom but not touch the basket or rack. For blanching by boiling, fill about ½ full, then begin heating.
6. Prepare the vegetable as recipe directs.
7. Blanch the vegetables, small amounts at a time, as each recipe directs.
8. Drain the vegetables well, then spread them in a single, even layer on a cookie sheet or sheets. Do not crowd vegetables on the sheet and do not prepare more vegetables than you can accommodate at one time.
9. For oven drying, put an oven thermometer toward the back of the tray. Put the tray on the top shelf in a preheated oven and maintain an oven temperature of about 120°F.
10. For box drying, turn on the light bulb to preheat the box. Place the tray on top of the box.
11. For both oven and box drying, check the trays often and stir the vegetables on the trays, moving outside pieces to center. For oven drying, turn the tray from front to back and, if making more than one tray, change the trays from shelf to shelf for even drying. Check the trays more frequently during last few hours of drying to prevent scorching.
12. When completely dried, as described in each recipe, remove the tray of vegetables from the oven or box and let it stand until cooled.
13. Turn the dried vegetables into a deep container, cover lightly with cheesecloth and condition, stirring once or twice a day, for several days.
14. Pack into airtight containers or double plastic bags and store in a cool, dark, dry place for up to 6 to 10 months.

15. To rehydrate, put the vegetables in a pan or bowl, pouring in just enough water to cover. Let them stand until soft, anywhere from a half hour to several hours, depending on the vegetable.
16. Cook vegetables in their soaking water until tender, or drain and add to recipes just as you would fresh vegetables.

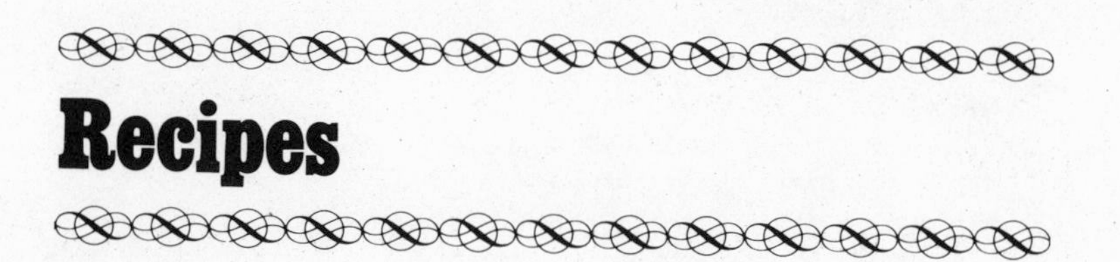

Recipes

CARROTS

1. Have ready a knife, vegetable peeler, blancher, cookie sheets, and an oven thermometer.
2. Choose crisp, young, tender carrots.
3. Wash well, peel and cut off tops and ends.
4. Slice very thin crosswise or lengthwise.
5. Blanch in boiling water about 5 minutes, in steam about 6 minutes. Drain well.
6. Arrange the slices in a single, even layer on cookie sheets.
7. Dry until very tough and leathery, usually around 12 hours.

Note: You may shred carrots for drying. Put shreds in a cheesecloth bag for blanching and cut the blanching times in half. The shreds will be done when dry and brittle.

CELERY

1. Have ready a knife, blancher, cookie sheets and oven thermometer (for oven drying).
2. Choose young, tender stalks with tender, green leaves.
3. Wash the stalks and leaves well; shake dry. Trim the ends.

4. Slice the stalks thinly.
5. Blanch in boiling water about 2 minutes, in steam about 3 minutes. Drain well.
6. Spread the slices in an even single layer on a cookie sheet. Spread leaves in an even, single layer.
7. Dry until brittle, usually around 12 hours.

Note: You can dry very tender, young spinach leaves or other greens as you do celery leaves.

CORN

Save the dried kernels to grind for your very own corn meal. A nut or seed grinder, special grain mill attachment for a mixer, or a heavy-duty blender can handle the grinding. Use corn meal for cornbread, breading, or spoon bread.

1. Have ready a blancher, knife, cookie sheets, oven thermometer (for oven drying), hammer.
2. Choose young, tender ears of very fresh corn.
3. Husk the ears, remove the silk and wash.
4. Blanch in boiling water about 5 minutes, in steam about 6 minutes.
5. Cut the kernels from ears and spread them in a single, even layer on cookie sheets.
6. Dry until very brittle, sometimes longer than 24 hours. Tap a single kernel with a hammer; if done, it will shatter easily.

MUSHROOMS

1. Have ready a knife, blancher, cookie sheets, oven thermometer (for oven drying).
2. Choose young, fresh, evenly sized, tender mushrooms with tightly closed heads.
3. Wash them very well, scrubbing with a brush. Remove and discard any stalks that are tough or woody, trim off the ends of any remaining stalks.
4. Slice or leave medium and small mushrooms whole, as you wish. Large mushrooms should be sliced.

(Continued On Next Page)

5. Blanch in boiling water for about 3 minutes, in steam for 4 minutes. Drain well.
6. Arrange in a single, even layer on cookie sheets.
7. Dry until leathery and hard, usually around 12 hours. Small pieces may be brittle.

ONIONS

There is no need to blanch onions, and no need to rehydrate them either, as long as pieces are small and you are adding them to other foods that have some moisture.

1. Have ready a knife, cookie sheets and oven thermometer (for oven drying).
2. Choose large, flavorful, perfect onions.
3. Cut off the stems and bottoms, remove the peel.
4. Slice the onions very thin or chop finely. Separate the slices into rings.
5. Arrange the pieces of rings in a single, even layer on cookie sheets.
6. Dry until very crisp and brittle, usually around 12 hours.

PEAS AND BEANS

Shelled green peas, black-eye peas, lima beans, pinto or soybeans all can be dried. Shell all beans before blanching, except soybeans.

1. Have ready a blancher, cookie sheets, oven thermometer (for oven drying) and a hammer.
2. Choose young, tender beans or peas.
3. Shell all peas or beans, except soybeans.
4. Blanch in boiling water about 3 minutes for green peas, 10 minutes for soybeans. Blanch in steam 5 minutes for green peas, 12 to 15 minutes for soybeans. Drain well. Shell the soybeans.
5. Arrange in a single, even layer on cookie sheets.
6. Dry until very crisp, usually more than 12 hours. Tap a single pea or bean with a hammer. When done, it will shatter easily.

SWEET PEPPERS

Peppers do not need to be blanched. Dried sweet peppers are handy to toss into soups, stews and casseroles.

1. Have ready a knife, cookie sheet and oven thermometer (for oven drying).
2. Choose tender, ready to eat, sweet red or green peppers.
3. Cut out the stem and seeds.
4. Chop in ¼-inch pieces.
5. Arrange in a single, even layer on cookie sheets.
6. Dry until brittle, usually around 12 hours.

PUMPKIN

Dry as for Winter Squash.

SQUASH, SUMMER

1. Have ready a knife, blancher, cookie sheet and oven thermometer (for oven drying).
2. Choose young, tender squash with tender skins.
3. Wash well, cut off the ends and slice about ¼ inch thick.
4. Blanch in boiling water about 3 minutes, in steam about 4 minutes. Drain well.
5. Arrange in a single, even layer on a cookie sheet.
6. Dry until brittle, usually around 12 hours.

SQUASH, WINTER

Peeling squash is not the easiest job in the world, but the storage space you save by drying squash makes it worthwhile.

1. Have ready a knife, vegetable peeler, blancher, cookie

(Continued On Next Page)

sheets and oven thermometer (for oven drying).
2. Choose mature, well-shaped squash.
3. Cut the squash into chunks, scrape out the seeds and string. Cut in 1 inch wide slices. Peel.
4. Slice 1-inch strips crosswise into thin slices.
5. Steam blanch about 10 to 15 minutes or until tender.
6. Arrange in a single, even layer on cookie sheets.
7. Dry until tough, usually more than 12 hours. Thinner slices may be brittle.

HERBS

Herbs need no pre-treatment before drying, just careful selection and gentle harvesting.

1. Have ready paper towels, cookie sheets or brown paper bags.
2. For herb leaves, choose herbs that are just about to blossom. Make sure the herbs are tender and well-colored, with perfect leaves and no bugs. Cut off the top $^2/_3$ of the plant. Pick early in the morning, if possible. For herb seeds, choose seeds that are fully developed and mature.
3. Wash the herb leaves if dusty or dirty. Shake them gently and pat dry with paper towels.
4. To Bag-Dry: Gather stalks in small handfuls and put them in medium-size brown paper bags. Hold the ends of the stalks at the top opening of the bag. Tie the bag's top around the ends of the stalks with string. Label the bag. Punch a few holes in bottom or side of the bag. Hang the bags from the string from hooks or hangers in an attic, on a covered porch or in any other warm, dry spot. Herbs are completely dry when the leaves fall from the stalks and can be easily crumbled between your fingers. You can strip the leaves from the stalks to crush or bottle whole, or just leave them in paper bags until ready to use. If all the leaves have not dried evenly, strip the leaves from the stalk and spread them out on a cookie sheet. Dry in 200°F oven for 30 minutes or until crumbly.

   To Tray-Dry: Pull leaves from the stalk or leave them on the stalk to dry. Spread them in a single, even layer on a cookie sheet and dry in a 120°F oven, or dry in a drying

box until the leaves are evenly dried and crumble easily between your fingers. Shake or pull leaves from stalks before storing them in airtight containers.
To Dry in Microwave Oven: Rinse the herb stalks and pat dry, then put 3 or 4 stalks between several thicknesses of paper towels. Microwave cook for 2 or 3 minutes or until the leaves crumble easily.
Seeds can be dried just as leaves. To remove the outer covering from dried seeds, just rub a small number of seeds between the palms of your hands and then shake them gently to let outer covering blow away.

FRUIT LEATHERS

Leathers are really a combination of fruit butters and drying. You spread thick fruit puree in an even layer, then dry it to roll up, store, then chew and enjoy. Fruit leathers are great for snacks, for camping or back packing. See the chart that follows for the amounts of sugar to use for different kinds of fruit.

1. Have ready a knife, large kettle, food mill (sieve or blender), cookie sheets, wire racks, waxed paper or plastic wrap, oven thermometer.
2. Choose fully ripe fruits (listed below). Fruits for leathers can be less than perfect. Wash them well and be sure to cut out and discard all bad portions.
3. Measure 2½ quarts of fruit into a large kettle.
4. Add the sugar in amounts given in the chart below and a small amount of water (¼ to ½ cup) to prevent sticking. Heat just until boiling. Remove from the heat.
5. Press sweetened fruit through a food mill or sieve, or blend several batches in a blender until smooth.
6. Puree should be about the consistency of apple butter. If thinner than that, return it to the kettle and simmer, stirring frequently, until it is the proper consistency. Cool slightly.
7. Lightly grease or butter the cookie sheets.
8. Evenly spread the puree about ¼-inch thick on the cookie sheets.
9. Dry until the leather is firm to the touch and will come off

(Continued On Next Page)

the cookie sheet in one piece when an edge is lifted, usually around 12 hours.

10. Put several wire racks together and slip the leather from the cookie sheets onto the racks to cool completely.
11. Sift just a small amount of cornstarch over one side of the leather, then flip and sift cornstarch over the other side. Cornstarch will help keep the leather from sticking to the waxed paper.
12. Tear off sheets of waxed paper or plastic wrap a little longer than the pieces of leather.
13. Put each piece of leather on a piece of waxed paper or plastic wrap and roll up the leather and its wrapping.
14. Wrap the rolls in more waxed paper or plastic wrap and seal.
15. Store in a cool, dark, dry place, or refrigerate or freeze.
16. Cut the leather into strips or slices to serve.

Sugar Charts

2½ quarts (5 pints) prepared fruit	**cups sugar**
Apples, chunked or quartered	*2/3 to ¾ cup*
Apricots, halved	*1 cup*
Strawberries, hulled and crushed	*2/3 cup*
Raspberries, whole	*1 cup*
Peaches and nectarines, sliced (peel peaches, but not nectarines)	*1 cup*
Plums or prune-plums, halved and pitted	*1 to 1¼ cups*

Add a small amount of water to prevent sticking, if the fruit does not have much moisture. A ¼-cup or ½-cup of water should be sufficient for low-moisture fruit, such as apples.

Step 5: After fruit has been prepared and cooked and sweetened, press it through a food mill or run in blender until smooth.

Step 7: Lightly grease or butter the cookie sheets while the fruit puree cools.

Step 8: Turn the puree out onto the buttered cookie sheet.

Step 8: Spread the puree about 1/4 inch thick on the cookie sheets.

Step 9: Dry the leather on a drying box or in a low oven for around 12 hours.

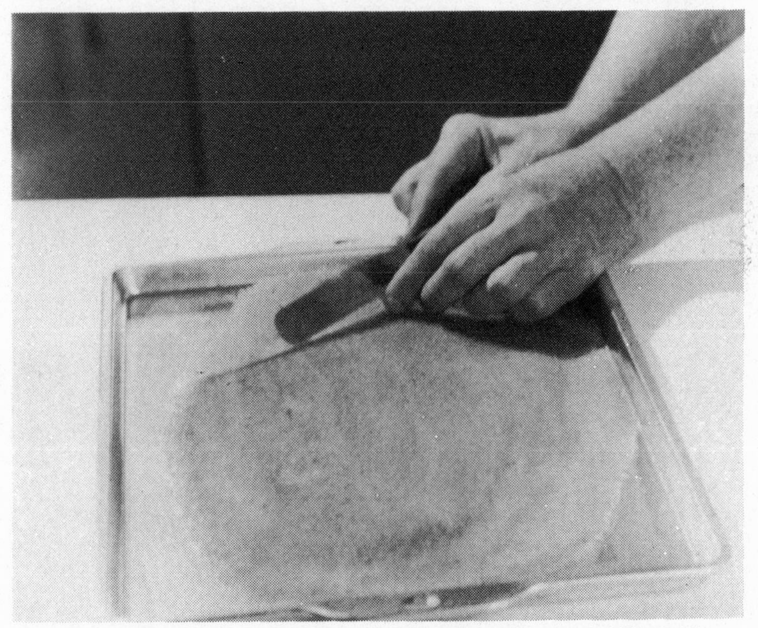

Step 9: Leather is done when it is firm to the touch and lifts off sheet in one piece.

Step 11: Sift a small amount of cornstarch on one side, then the other side.

Step 13: Roll up leather between two pieces of waxed paper for storage.

Smoking

The most primitive meat preservation processes used by the American Indians and the Colonists were sun-drying and smoking. Salting and brining were also used to preserve meats; salting meat in kegs was a common practice until crocks became widely available.

In rural America the smokehouse used to be a normal part

of established homes. In the South, the art of smoking hams and bacons became a glorious tradition. But the problem facing a farm family in any part of the country was not how to make a single tasty smoked shoulder, but how to effectively preserve the whole pig. The larger parts — shoulders, hams, and sidemeat — were cured in brine or salt for several weeks before hanging in the smokehouse. The trimmings and leftover meat scrapes were ground into sausages, then "fried down" — along with the spareribs and backbones — before being packed into crocks and sealed with layers of rendered lard.

Pemmican and jerky were the travelers' "instant" dried meats. They were lightweight, nourishing and kept well. Pemmican was originally made by Northern Indian tribes. First they dried thin-sliced meats over a fire— or just left them in the sun. Then the slices were shredded up and mixed with an equal portion of rendered fat, some marrow and a handful of wild cherries, or later, currants. Sewn up in rawhide pouches and sealed with tallow, pemmican was ready to go with the adventurer charting unexplored territory.

You will find a recipe for jerky at the end of this chapter. The original jerky was probably *charqui* from South America. Making *charqui* involved slicing meat and putting the slices in brine or salt and wrapping them in animal hide to cure before hanging in the sun to dry. Our updated recipe for jerky calls for thin-sliced meat to be cold-smoked. Like the South American original recipe, it is a good food to take on hikes or backpacking trips. And, unlike the modern commercially-made version, homemade jerky has no additives or preservatives.

Modern smoking, like ancient smoking, preserves meats and fish by drying and by long exposure to smoke from aromatic woods. The smoked meats, fish and turkey in this chapter are perfect for banquets or picnics. The simplified, basic directions for building a smoker, building a fire and preparing the food will bring back the rich but delicately-flavored smoked foods that disappeared from our tables when canning and freezing came of age.

Cold-Smoking

Smoking at temperatures below 120°F is known as cold-smoking. Cold-smoking preserves foods, but does not cook them. All cold-smoked foods can be stored, for varying lengths of time (depending on the food itself), in cool places or in the refrigerator or freezer. For safety's sake, we recommend refrigerator storage. Cold-smoked foods must be cooked before eating.

Hot-Smoking

Smoking food at temperatures above 120°F, usually 225°F to 250°F, is hot-smoking, or smoke-cooking. You can smoke-cook in a smoker or smoke house, as well as in a covered barbecue grill, a Japanese or Chinese smoke oven or one of the new little portable smoke cookers. Smoke-cooked foods are cooked and ready to eat. You handle and store them as you would any other cooked meat. Follow manufacturer's directions for smoke cookers.

Most of our recipes are a combination of cold-smoking and hot-smoking, starting off at low temperatures to build up plenty of smoke flavor, then finishing off at higher temperatures to cook food completely, and to save you from tending the fire through the night.

Curing and Brining

Most meats that are to be completely cold-smoked are treated before smoking with a brine (salt water) or cured (packed in dry salt or sugar). Curing hams and bacon requires special ingredients and equipment. We recommend that you get information and advice from your state cooperative extension office before embarking on large scale curing and cold-smoking projects. Keep in mind that cold-smoked, cured meat is still raw and must be cooked before you eat it.

Fire and Smoke

Use only hard woods for heat and smoke. Do not use evergreen tree wood, such as pine. Pine and other evergreen woods are resinous and will discolor and give the meat an off-flavor. Select hickory, apple, cherry, peach, oak or other hardwoods or fruit woods, or use dried corn cobs. Green wood is best for fire building; it burns and smokes with great amounts of heat, but without flames. If you cannot get green wood, add plenty of soaked chips or sawdust to provide dense, heavy smoke.

Start the fire with kindling and chips. Paper used as kindling will make an ash that will get on the food. Lighter fluid will leave an odor and taste in your smoked foods.

Arrange green hardwood logs spoke fashion.

Some expert smokers prefer to use no fire at all, just an old iron pan full of damp chips or sawdust set on an electric hot plate. The fire must burn low and even, and for many hours. If smoking must go on for more than a day, you will need to get up at night to check the fire and keep it going. When outside temperatures are cool, but not freezing, you can let the fire go out at night and start it again the next morning.

Temperature

Cold-smoking should be done at temperatures less than 120°F, usually 70°F to 90°F. Smoke-cooking temperatures are around 250°F, although you can smoke-flavor even at broiling temperatures by tossing a handful of dampened wood or chips onto the coals for the last 15 minutes of cooking time.

Use an oven thermometer inside the smoker to keep track of temperatures. Use a meat thermometer in large pieces of meat to be smoke-cooked then you will not have to guess when the meat is done.

The Smoker

All smokers have a few basic requirements. They need heat and wood to create the smoke; a cavity or enclosure to contain smoke; some arrangement to hold food in the smoke; and some drafting to assist the fire and help the smoke flow

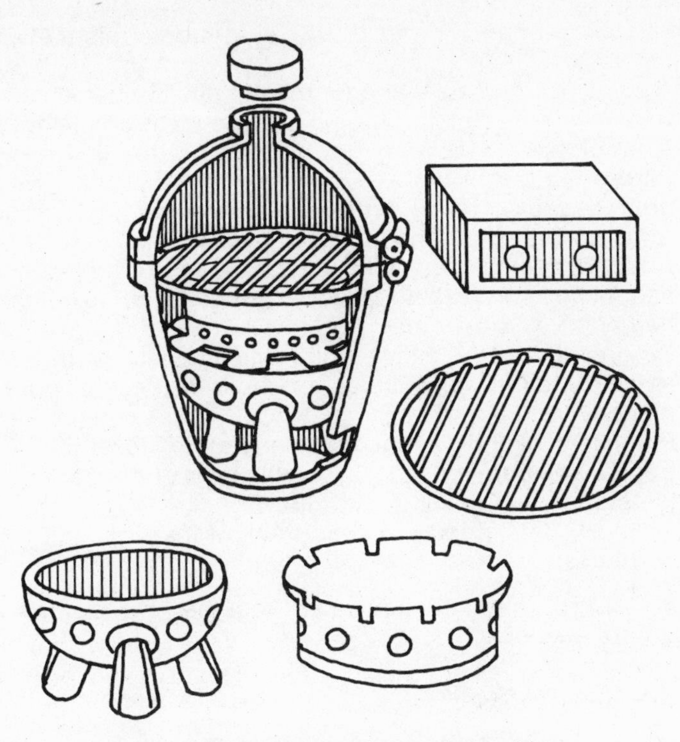

Japanese smoke oven for smoke-cooking.

around the food.

You can meet these requirements with a set-up as simple as a cardboard box put over a fire of smoldering twigs, or you can get as fancy as a custom-built cement block house with vents, fans and electrical heat source.

Our directions are for a relatively easy to put together barrel smoker, and a slightly more complicated refrigerator variation. For directions for more sophisticated smokers, we refer you to books on smoking, or your favorite lumberyard.

Barrel Smoker Materials and Equipment

1. A clean, 55 gallon metal barrel or drum, with ends removed. But, save the ends for covers, if you can get them.
2. A child-free area of your back yard, preferably with a slight slope going up in the direction of the prevailing

winds. You can use dirt or gravel to create this slope, if necessary.

3. A cover for the barrel and for the fire pit. The barrel cover can be wood or metal; the pit cover should be metal.
4. A broomstick or two, or dowels or pipe that will just fit across the barrel top and something (a cleat or wire) to hold the stick in place. You will also need several S-hooks to go on the stick. The meat hangs from the S-hooks.
5. A rack or grill that will fit inside the barrel to hold food that cannot be hung (fish fillets, for example). A circle cut from hardware cloth will also work.
6. Several three-inch stove bolts and nuts to go through holes drilled in barrel sides and hold the rack or grill in place.
7. Planks, plywood or sheets of metal to cover the fire trench. You can line the trench with terra cotta, cement or metal pipe and cover it with dirt.
8. An oven thermometer to keep track of the smoker's temperatures.

Now that you have the equipment together, here is how to turn it into a smoker.

Making a Barrel Smoker

1. Drill several holes through the sides of the barrel about a third of the way down from the top. Put in the stove bolts

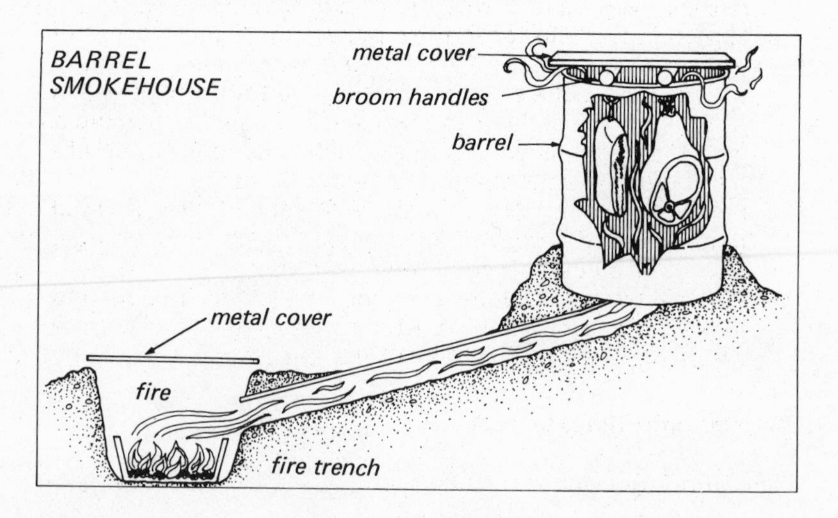

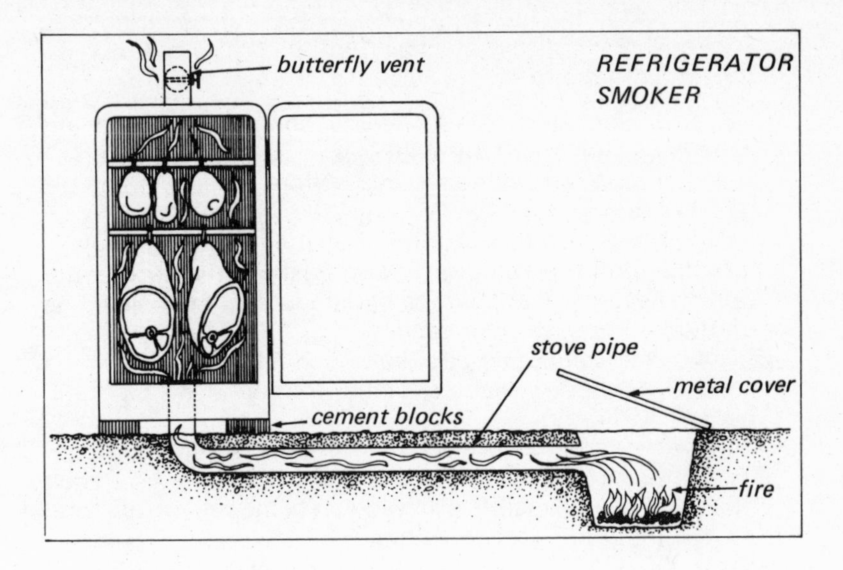

and fasten with nuts to hold them in place. Put the rack, grill or hardware cloth in place, resting on the stove bolts. (Remove the rack when smoking food that can hang from S-hooks.)

2. Put the S-hooks on the broomsticks, then fasten the sticks in place across the top of the barrel.
3. Select a site for the barrel on slight natural rise or on a rise that you have created from dirt, sand or gravel.
4. Now select a site for the fire pit, about 10 feet down hill and down wind of the barrel. The upward slope and wind will help move the smoke from the pit to the barrel. Dig the pit about two feet deep and two feet in diameter, then dig a trench six-inches wide and six-inches deep from the fire pit to directly underneath the center of the barrel.
5. Line the trench, if you wish, with pipe. Cover a lined trench with dirt; cover an unlined trench with boards, planks or pieces of metal, then with dirt.
6. Put the barrel in place over the end of the trench and mound dirt around the barrel to keep the smoke from escaping, but be sure to keep the trench free.

Refrigerator Smoker

An old refrigerator can make a dandy smoker (if your neighbors don't mind the looks). It already has shelves for holding or hanging food. Buy a chain and padlock to keep the refrigerator shut and empty when you are not around.

1. Remove the compressor, freezer compartment and any plastic or nonmetal parts inside.
2. Cut a round hole in the top and bottom, so smoke can get in and leave.
3. Seal ducts or openings other than those you have cut.
4. Put a butterfly draft in the hole in the top of the refrigerator and select an elbow section of pipe to fit in the hole in the bottom of the refrigerator.
5. Choose the site and build the trench and fire pit as in the Barrel Smoker directions. Line the trench with pipe and cover it with dirt.
6. Put two cement blocks on either side of the trench opening for the refrigerator to stand on. Connect the trench pipe to the elbow pipe in the bottom of the refrigerator and stand the refrigerator in place.

Basic Steps for Smoking

1. Start the fire in the pit, using sticks or chunks of green hard wood. When the fire is going well, add a handful of dampened chips or sawdust to get plenty of smoke.
2. Put an oven thermometer in place in the smoker on a rack or hanging from one of the S-hooks. Cover or close the smoker.
3. Prepare food as the recipe directs. If you are smoking large cuts of meat, insert a meat thermometer in the center of the largest muscle, with the tip away from bone and fat.
4. When the smoke is flowing well and the inside temperature is about 85°F, add the food.
5. Smoke the food, checking the fire frequently, adding wood and adjusting the barrel and pit covers to keep the fire burning evenly and the smoke flowing. Check the thermometer once or twice during smoking to be sure the desired temperature is being maintained.
6. Add more wood and increase the drafts if the recipe directs you to increase the temperature.
7. Check the meat thermometer or recipe directions to determine when the food is done and ready to be removed.

Food Storage

Cold-smoked foods can be refrigerator-stored for several

weeks, or freezer-stored for up to six months. Be sure to wrap the food in moisture-vapor proof material and seal carefully. Smoke-cooked foods should be eaten at once or chilled to be eaten within the next few days.

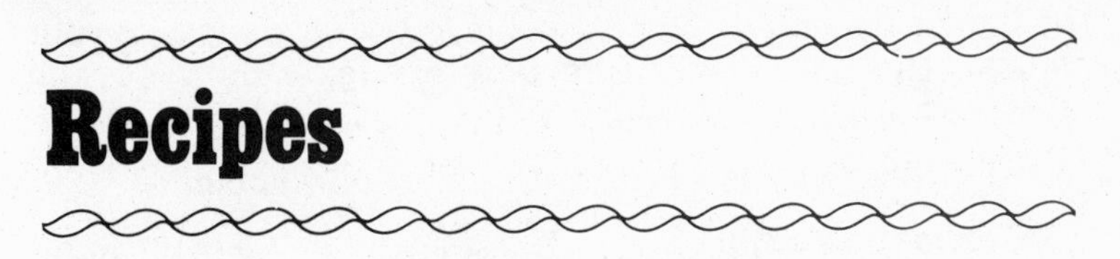

Recipes

SMOKED CIDERED PORK CHOPS

You start off with cold smoking, to develop plenty of great flavor, then increase heat to complete cooking. Try ribs this way, too.

Ingredients

4 to 6 lean pork loin or shoulder chops, ¾ to 1 inch thick or 4 to 6 pounds spareribs
1 pint cider
1 teaspoon instant minced onion
½ to 1 teaspoon ground cinnamon or nutmeg

Equipment

Large heavy-duty plastic bag
Smoker with racks or covered barbecue grill

1. Put the chops or ribs in the plastic bag. Pour in the cider; add onion and spices.
2. Refrigerate several hours or overnight.
3. Lift the meat from the marinade and arrange it on racks in the smoker or covered grill.
4. Smoke at 85°F with plenty of smoke for 1½ hours.
5. Increase the temperature to 225°F to 250°F and keep up plenty of smoke for another 45 minutes to 1 hour, or until meat is very tender and falls from the bones.
5. Brush the meat with any leftover marinade during last 15 minutes of cooking time, if desired.

NOTE: To smoke-cook in a covered barbecue grill, follow the manufacturer's directions.

SMOKED FISH FILLETS

You can easily double this recipe, if you put two racks in your smoker. The herbs listed are only a suggestion, you can choose your own seasoning.

Ingredients

1 gallon cold water
1 cup salt
5 to 10 pounds fillets of firm, white-fleshed fish, either fresh or salt water fish
Any of the following:
3 bay leaves
1 tablespoon dill seed
½ teaspoon white pepper
1 teaspoon onion or garlic salt
1 teaspoon marjoram or tarragon
1 teaspoon rosemary
Oil

Equipment

Large mixing bowl or pan
1 quart measure
1 cup dry measure
Spoon
Spatula or turner
Smoker with racks

1. Stir together the water, salt and choice of herbs until the salt dissolves.
2. Arrange the fish fillets in a mixing bowl or other large container (such as a plastic bag, clean plastic dishpan, glass baking dish).
3. Pour the brine over the fish. Cover and refrigerate several hours or overnight.
4. Drain the fish well, then rinse with cold water.
5. Spread out the fillets on racks and let them dry completely. Use a fan to hasten the drying, if desired.
6. Arrange the dried fillets on a greased rack in the smoker.
7. *Cold-smoking.* Smoke at 70°F or less for four or five days, keeping the fire and smoke going all the time.
8. *Smoke-cooking.* Smoke by starting the fish in the smoker at 90°F for 20 minutes, then increasing temperature to 140°F, until the fish flakes with a fork, about 1 or 2 hours. Smoke-cooked fish will be browned, fully-cooked and ready to eat.

FISH THAT SMOKE WELL. Freshwater: walleye or Northern pike, bass, catfish, perch, whitefish, trout.
Saltwater: pompano, cod, flounder, whitefish, halibut, pollock, mullet, mackerel.

SMOKED SHRIMP

Smoked shrimp are among the most elegant appetizers or main dishes you can serve.

Ingredients

- 10 *pounds peeled raw shrimp*
- 1 *quart water*
- ½ *cup salt*
- ½ *cup sugar*
- 1 *tablespoon mixed pickling spice or 1 packet prepared spices for crab and shrimp or any of the spices listed under Smoked Fish Fillets*
- *Oil*

Equipment

- *Large mixing bowl or pan*
- *1 quart measure*
- *1 cup dry measure*
- *Spoon*
- *Paper towels*
- *Wire rack*
- *Smoker with racks*

1. Prepare a brine from water, salt, sugar and spices as directed in recipe for Smoked Fish Fillets.
2. Refrigerate the shrimp in brine 30 minutes to 2 hours, depending on the strength of flavor you prefer.
3. Drain the shrimp and rinse well with cold water.
4. Pat dry with paper towels, then arrange them on a wire rack and let them air dry about 30 minutes.
5. Oil or grease the racks in your smoker and arrange the shrimp on the racks with space between for smoke to circulate. Use a wire cake rack if the spaces in the smoker's rack are too wide.
6. Start smoking at 90°F. After 15 minutes gradually increase the temperature to 135°F to 140°F and smoke until the shrimp are done, about 1 to 1½ hours. Taste the shrimp to check readiness and smoke for a longer time, if necessary.

SMOKE-FLAVORED SALT

Smoke some salt while you are smoking other food; then store salt so you can add smoke flavor any time you want it.

Ingredients	**Equipment**
Table salt	*Shallow pan* *Smoker with rack*

1. Spread the salt in shallow pan.
2. Put the pan on rack in the smoker while you are smoking other foods at 85°F or so.
3. Smoke about 3 to 4 hours or until golden in color.

SMOKED OYSTERS

Serve these succulent morsels as appetizers at your next big party; when your guests rave about them casually mention that you smoked them yourself.

Ingredients	**Equipment**
Oysters, cleaned, shucked and washed	*Large mixing bowl or pan*
1 *gallon water*	*1 quart measure*
1¼ *pounds salt*	*Smoker with racks*
Salad oil	

1. Mix the water and salt until salt dissolves.
2. Pour brine over the oysters and let stand 5 minutes.
3. Drain the oysters thoroughly.
4. Pour just enough oil over the oysters to lightly coat them.
5. Arrange them on a well-greased rack in your smoker with space between oysters for air to circulate. (Use a wire cake rack if the spaces in the smoker's rack are too wide.)
6. Smoke at 180°F for 15 minutes; turn and smoke 15 minutes longer.
7. Eat the oysters right away. Or refrigerate and eat them within a few days. Oysters can be packaged in freezer containers and frozen for up to 2 months.

JERKY

You may add almost any spices or herbs you wish to the basic jerky marinade. Remember that very spicy or salty jerky is better for at-home eating, where you can easily quench your thirst. For camping or backpacking, stick to mild seasonings.

Ingredients

3 pounds lean round, flank or chuck steak
½ cup soy sauce
2 tablespoons sherry (optional)
2 tablespoons vinegar (optional)
½ teaspoon powdered ginger, garlic powder, or onion powder

Equipment

Sharp knife
Mixing bowl or heavy duty plastic bag
Measuring spoons
Smoker with racks or covered barbecue grill

1. Freeze the meat until it is just beginning to get firm, then slice it across the grain into ⅛ to ¼-inch strips. Cut away and discard all fat and connective tissue.
2. Combine the soy sauce and any other ingredients you wish to use (just soy sauce is enough) in a large mixing bowl or heavy-duty plastic bag.
3. Add the meat strips and mix well to coat each strip. Cover and refrigerate several hours or overnight.
4. Lift the meat from the marinade and drain well.
5. Arrange the meat strips on racks in a smoker.
6. Smoke at 90°F to 110°F about 10 to 12 hours, or until the strips are dry and brittle.
7. Store the jerky in plastic bags or airtight containers.

NOTE: To smoke in a covered barbecue, add a handful of dampened chips to very low coals that are left after you have barbecued other foods. Arrange the meat strips on the grill, cover and smoke as above.

SMOKED BEEF, LAMB OR PORK ROASTS

Select top-quality meats and cuts that you would normally oven-roast. Smoking time depends on the size of the roast.

Ingredients	Equipment
Beef, lamb or pork roast	*Smoker with racks*
Seasoned salt	*Meat thermometer*

1. Preheat your smoker to 225°F.
2. Rub the meat well with seasoned salt or your own blend of salt and spices.
3. Insert a meat thermometer so the tip is in the center of the thickest muscle, away from fat and bone.
4. Smoke until done to your taste. Pork must reach an internal temperature of 170°F.

NOTE: To smoke-cook in a covered barbecue grill, follow the manufacturer's directions.

TWO-DAY SMOKED TURKEY

Using a combination of cold and hot smoking saves you from more than one night of fire-tending.

Ingredients	Equipment
1 (10-pound) turkey	*Heavy duty plastic bag or other large container*
1 gallon cold water	*Smoker with racks or covered barbecue grill*
1 cup salt	*Meat thermometer*
2 teaspoons rosemary leaves, crushed	
Sherry or white wine	

1. Rinse the turkey inside and out with cold water.
2. Combine the cold water, salt and rosemary. Stir until the salt dissolves.

3. Put the turkey in a large plastic bag or mixing bowl and pour the brine over it.
4. Refrigerate the bird for several hours or overnight, turning it in the brine occasionally.
5. Lift the turkey out of the brine, rinse it well and pat dry with paper towels. Let it stand in a cool, drafty place (in front of fan, if you wish) until the skin is dry. Put a meat thermometer in the thickest part of the thigh with the tip away from the bone.
6. Smoke at about 100°F with plenty of smoke for 10 hours.
7. Increase the temperature to 225°F to 250°F and keep up plenty of smoke for another 8 to 10 hours, or until the meat thermometer reaches 185°F.
8. Brush with wine during last hour of cooking, if desired.
9. Serve immediately or refrigerate and serve cold.

NOTE: To smoke-cook in a covered barbecue grill, follow manufacturer's directions.

Cold-Storage

The colonists cooked on open hearths — no thermometers, no clocks, no standardized measuring equipment, and no refrigerators. Almost everything had to be produced at home and stored at home — no gargantuan central cold-storage areas that we call supermarkets. During the early years, any extra harvest was stored in cool caves or tunneled-out, straw-lined pits that could withstand freezing temperatures. Estates in the South often had separate out-buildings for different kinds of cold storage. Later, houses were built with root cellars or cold, damp basements intended as cold-storage areas.

These chilly spots were perfect places to store root vegetables, celery, pumpkin, squash, apples, even pears, to keep their goodness as long as possible after harvest. Our warm and dry basements or utility rooms have replaced cellars and, while we have gained comfort, we have lost storage space. All these foods can be stored in the refrigerator — but that space is valuable and badly needed for short term storage.

As a result, you will need to plan, and perhaps construct, a special spot for cold storage of your garden's bounty. We recommend cold storage only for someone who plans and plants a big garden. Buying produce from a supermarket to hold in cold storage just does not make sense — let the market do the cold storing for you.

What you plan and construct depends on where you live. In milder climates, where frost is infrequent and does not penetrate too deeply, a barrel or well-insulated mound will keep vegetables. In colder areas you will need to take extra precautions against freezing. If you live where it is always warm, just forget about cold storage — it is not possible. Your state cooperative extension office will have information appropriate for your area about types of cold storage, building plans for root cellars, even suggestions about which varieties of fruit and vegetables to plant.

Foods For Cold Storage

Never try to store damaged or imperfect produce. Late ripening and maturing fruits and vegetables are the best candidates for cold storage. Certain varieties take better to this type of storage than others, so check seed catalogs and packets before you buy and plant. Carrots, beets, turnips, parsnips, potatoes, and winter squash are wise choices for cold storage from the vegetable family, though potatoes need special handling. Apples are the best fruits to store this way. Some of the firm varieties of pears can be held in cold storage, too.

Handling

Harvest fruits and vegetables as late as possible. For many vegetables, this means planting them later than usual for a late harvest. Wait until the first frost warnings to harvest. Carrots, parsnips, potatoes and turnips, for example, can stay in the ground even after the first frost or two, if the ground is well mulched. Pick fruits and vegetables and handle them carefully to avoid bruising. Use only perfect fruits or vegetables for cold storage. One bad item can spread decay to others and ruin the whole box, barrel or mound.

Harvest on a dry day, if possible, and let the vegetables dry on the ground, in the sun, for several hours before packing them away. Onions often need several days of drying; potatoes, however, should not be exposed to sun or wind damage. Produce should be cool when packed.

Wash vegetables and fruits, if you must, but most experts agree that all you really need to do is brush off excess dirt. Produce should be dry before it is packed.

Curing

Potatoes, pumpkins and most squash have to be cured before storing. Curing means holding the vegetables at a warm temperature (70°F to 85°F, depending on the item) for a week or two. Curing hardens the skins and rinds and helps heal surface cuts, reducing mold and rot damage.

Packing

Some vegetables and fruit (potatoes, onions, squash, apples) can go right into cardboard boxes, barrels, polyethylene bags or other containers. The root vegetables (beets, carrots, turnips, parsnips) are better packed in some type of material to insulate them and keep them from touching each other, so decay will not spread. You can wrap each vegetable in newspaper, then pack them loosely in boxes, barrels or plastic bags. If you use polyethylene bags, poke a few holes in the bags so there will be some ventilation. Purists prefer more natural packing materials, such as damp or dry sand, sawdust, peat, sphagnum moss, leaves, straw, or shavings. Line the container with a layer of packing material, then arrange a layer of vegetables, with space between each. Fill in around

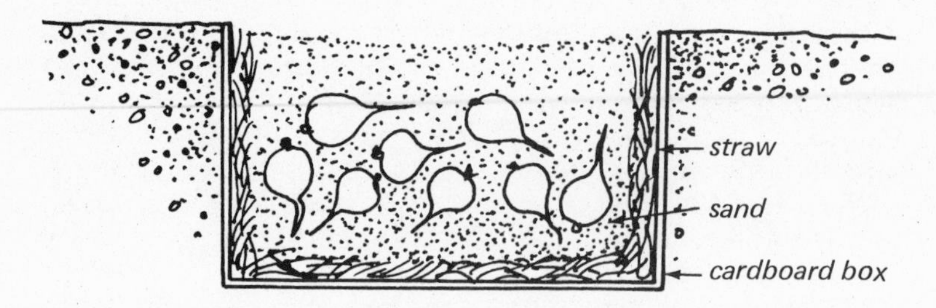

BOX STORAGE

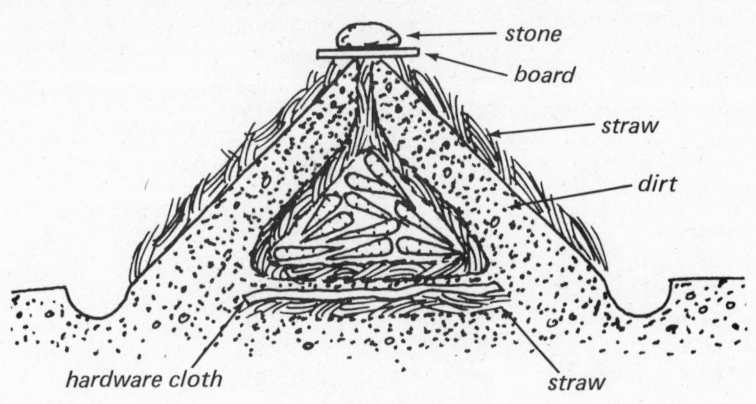

MOUND STORAGE

each vegetable and on top with another layer of packing material. Repeat these steps until the container is full.

Outside Storage

Where winters are mild and there is not much snow, outside storage is an easy answer to holding large crops. You can store fruits or vegetables (separately, never together) in a mound or in a barrel.

Mound Storage

Plan to dig separate pits for fruits and vegetables. Locate the mound in an area with good drainage. Dig a shallow dish-shaped hole six to eight inches deep in your garden and line it with straw or leaves. Spread the straw bed with a sheet of hardware cloth or screen (to keep out burrowing animals) and then stack fruits or vegetables in a cone on the prepared bed. Separate individual piece and layers of produce with packing material. Making a volcano shape, cover the mound with more straw or leaves, then shovel on three or four inches of dirt. Cover all but the top of the "volcano." Pack the dirt firmly with the back of your shovel. Pile on another thick (six- to eight-inch) layer of straw. Do not cover the top of the volcano, but leave it open for ventilation. Put a piece of board on top of each mound to protect it from the weather. You may have to weight the board with stone or brick to keep it in place. Dig a

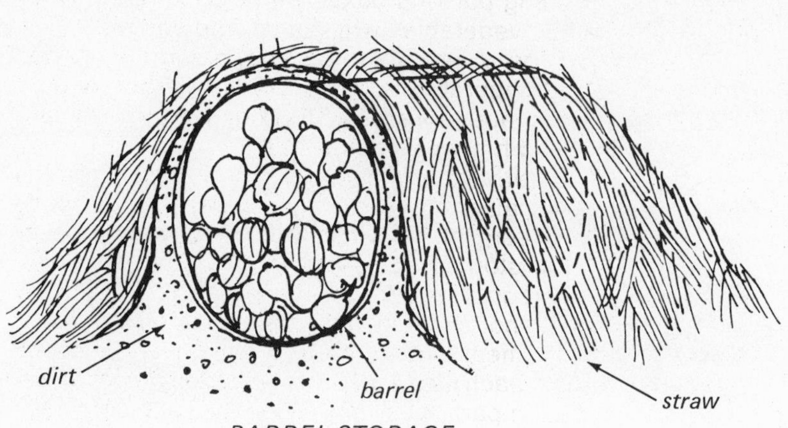

BARREL STORAGE

shallow drainage ditch around the mound.

When you open a mound you probably will have to bring all the fruit or vegetables in the house, especially if the ground has frozen. Repacking is almost impossible. Several small mounds, each containing a bushel or so, are more practical than one large mound. You can store several kinds of vegetables in the same mound if they are separated by packing material — that way you can enjoy a bushel of mixed vegetables instead of all carrots or all potatoes. The U.S. Department of Agriculture recommends changing the location of the pits every year to avoid contamination.

Barrel Storage

Nest a wooden barrel in a straw-lined hole, pack in the vegetables and then cover the barrel with several insulating layers of straw and dirt. Be sure to mark where the mouth of the barrel is, so you can find it easily when you are ready to dig out vegetables.

Indoor Storage

Storing vegetables indoors requires low temperatures, ventilation and some type of humidity control. These requirements do not call for fancy equipment: humidity can be maintained by putting pans of water on the floor, covering the floor with damp straw, sand or sawdust, using damp sand or sawdust for packing, or lining packing boxes with plastic bags. Without extra humidity, vegetables will dry up and shrivel.

Ventilation can be a vent to the outside, a slightly opened door or window. Temperatures, with a few exceptions, should be below 40°F and above freezing. Potatoes, pumpkins and winter squash need higher temperatures.

Your house (or perhaps another building on your property) probably has several possible storage areas. If your house has an outside basement entrance with stairs going down, you can use it as a storage area — the stairs can be shelves. You will need a door at the top of the stairs, and probably another door at the bottom of the stairs, over the existing one, to hold in the basement's heat. Use a thermometer to check the temperature on each step and put food where the temperature is right for each particular item. Window wells can make nifty little storage areas, if they do not collect and hold water. Line the wells with straw or bedding, put in the vegetables and

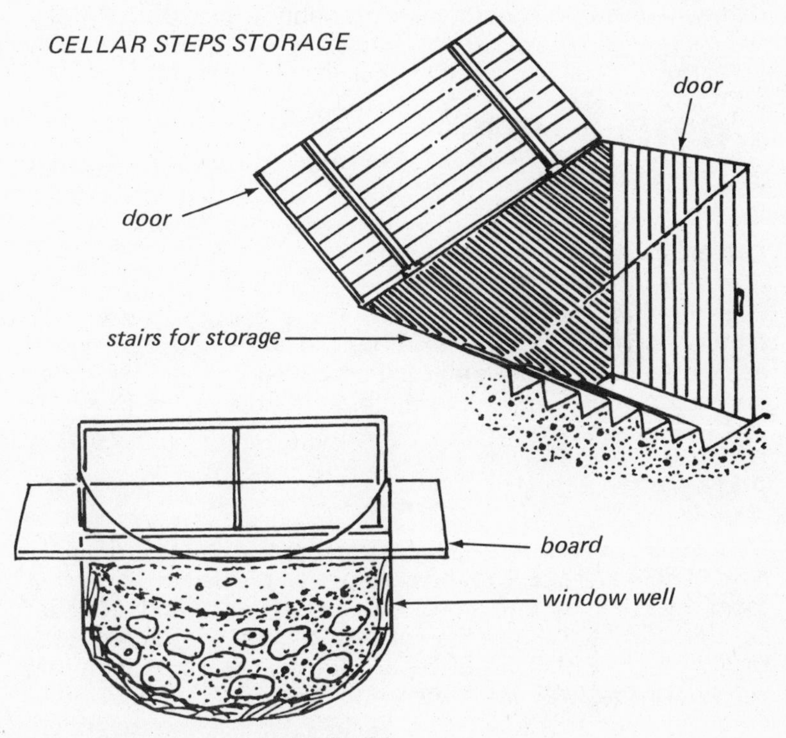

add packing material, if necessary. Then cover the wells with boards or more bedding. If the windows open inwards you may be able to take vegetables out from the basement, and not even have to go outside and dig!

Look around your property to see if there is some arrangement like these examples. Use a thermometer to test temperatures and be sure to allow for humidity and just a little ventilation.

You can go all out and build a cold storage room in your basement. What you will be doing is creating a separate little room, insulated from heat. Gardening magazines, agricultural extension offices and lumberyards all have plans for constructing indoor cold storage rooms.

Basically, you will have to partition off an area without heating pipes or ducts. The area can be divided into separate areas for fruits and vegetables. For ventilation, there should be a window — two or more windows if the room is partitioned. For air circulation, plan to have removable slatted flooring and shelves. Slatted flooring makes it easier to use dampened sawdust or other wet material to raise the humidity.

Basic Steps

ONIONS

Dig the onions and leave them on the ground to dry completely, usually about a week. Pack loosely, without any packing materials, in well-ventilated containers. If you like, braid the tops together and hang the onions from hooks in a cold storage area. Store at 32°F to 40°F.

POTATOES

Choose late maturing varieties. Early potatoes are difficult to hold in cold storage. Dig the potatoes when the soil is dry. Let them dry briefly before storing. Avoid sun and wind damage. Cure by storing them at regular basement temperatures (60°F to 70°F) in moist air for a week or two. Then pack them into boxes or other well ventilated containers, but without additional packing material. Store at 45°F to 50°F in moderately moist air.

PUMPKIN AND WINTER SQUASH

Harvest just before the first frost, leaving an inch or two of stem. Cure all but acorn squash at 80°F to 85°F for ten days, or for two to three weeks at lower temperatures. After curing, move them to a cooler spot for longer storage. Store in a dry place at 55°F for 6 to 8 weeks. Acorn squash can be kept in a dry place at 45°F to 50°F for over a month.

ROOT VEGETABLES

Choose late maturing varieties of carrots, beets, celeriac, kohlrabi, rutabagas, turnips. Leave them in the ground until frosts begin. Dig them up when soil is dry and leave them on the ground to dry for three or four hours. Cut off the tops, leaving two or three inches of stem. Pack in packing material in boxes, barrels, polyethylene bags with air holes, or in a mound. Store at 32°F to 40°F.

SWEET POTATOES

Put sweet potatoes directly into storage containers when you harvest them. Like white potatoes, sweet potatoes have to be cured. They cure best under moist conditions at 80°F to 85°F for 10 days. At lower temperatures, curing takes longer — two

to three weeks. Stack storage crates and cover them to hold in the humidity while curing. After curing, store the sweet potatoes at 55°F to 60°F. Storage at 50°F or below will damage sweet potatoes. Outdoor pits are not recommended for storage. With proper procedures, sweet potatoes can be stored through the winter.

TOMATOES

Plant late so the vines will still be vigorous when you pick the tomatoes for storage. Harvest them just before the first killing frost. When you harvest, remove the stems from the tomatoes, then wash and dry them before storing. Be careful not to break or scratch their skins.

Separate the green tomatoes from those that are showing red. Ripen green tomatoes by packing them one or two layers deep in boxes or trays. At 65°F to 70°F, mature green tomatoes will ripen in two weeks. At 55°F ripening will be slowed down to nearly a month. Immature green tomatoes will take longer to ripen; tomatoes showing red cannot be held in storage as long as green ones.

Tomatoes in storage need a moderate amount of humidity, good air circulation and a temperature of 55°F to 58°F. Check your tomatoes once a week to remove the ripe ones and any that have decayed.

APPLES

Select varieties that are known to be good keepers. Ask your local agricultural extension office for a list of apples for cold storage. Pick the apples when they are mature but still hard. Pack them in barrels or boxes, but without any additional packing material. Store at 32°F and be sure to arrange for a moderate amount of humidity.

An extra refrigerator in the basement is an easy way to store apples.

Freezing

The iceman with his horse and wagon have gone the way of the pickle barrel, the pantry and the smokehouse. Indeed, mechanical freezing and cooling have transformed the kitchen. The heart of the home has been moved from the hearth or range to the refrigerator-freezer. Today's children will conjure up memories of security and home when they recollect standing in front of their well-stocked freezer and refrigerator contemplating which goody to sample.

Of course, freezing foods to preserve them is nothing new — Eskimos have done it for centuries. Early American farmers, fishermen, and hunters in cold climates knew all about hanging fish and game out of the reach of animals to freeze all winter.

America's passion for ice cream and iced drinks started early and influenced the development of the freezer. In the colonial South, plantation icehouses could store enormous quantities of ice — enough for ice puddings, iced beverages and ice cream to last all summer. Ice cream was one of the main motivations for icehouses in northern communities, too. The village icehouse was stocked with frozen pond water which was sawed into cakes. Ice production became a major New England industry after the first decade of the 19th century. Large icehouses shipped ice to cities far to the south — even to the West Indies.

An ice-making machine, invented in 1851, was gradually perfected and by the end of World War I, replaced natural icehouses. The icebox, invented to cool milk, became a standard kitchen appliance by the latter half of the 19th century, followed by the electric refrigerator in the 1920s. But the major step towards frozen convenience foods was taken by Clarence Birdseye. Without his quick-freezing method, frozen packaged goods would not have become popular. Once the new fast-frozen foods were accepted, proper storage units were quickly developed for grocery stores and for the home. The golden age of the freezer did not dawn until the late 1930s.

Modern Freezing Methods

Freezing is a very simple method of food preservation. It takes only a few steps: selection of good food, washing, preparing, blanching, chilling and packaging. Then coldness does the rest of the work. And we do mean cold: 0°F.

Zero temperatures stop the growth of microorganisms that can cause spoilage. Cold does not kill them, as heat preservation does, but it stops them while food is frozen.

Although freezing procedures are simple, they should be followed exactly. Select perfect food, clean and handle it with care, package and seal it properly and keep it in a well-managed freezer and you will be able to take out the same quality of food.

Convenience and flavor are the big advantages for freezing. It is easy to prepare food for the freezer and it is easy to prepare food for the table from the freezer. Fruits and vegetables from the freezer taste more like fresh than their canned counterparts. But, freezing is a more expensive form of storage than canning. The freezer itself is an investment and it takes electricity. However, if your time is valuable, and if you manage your freezer wisely, it can help you save on food costs.

Freezer Management

For the most efficient use of your freezer you must be organized. Think of your freezer as a warehouse or a food depository. You need to keep track of what is inside, when it went in and when it should come out. "First in, first out" is the byword for the best in flavor and appearance in frozen foods. The sections that follow tell you the maximum storage times for specific food groups. The food is still safe to eat after these maximum times, but may not be at the peak of its quality. As a rule of thumb, rotate your stock about every 6 months, or freeze only enough fruits and vegetables to last until the next season.

Take care of your freezer just as the use and care instructions that came with it tell you. Keep the freezer defrosted, free of ice and clean — it will work better and cost you less to operate. A full or almost full freezer is cheaper to run than an empty or almost empty one. If your freezer needs an annual or semi-annual defrost, do it while the weather is cold, preferably before you start planning your garden. During a defrosting in cold weather, not only can the food wait outside (in well-insulated boxes or coolers) but you can take a thorough inventory and then determine how much to plant for freezing, or how much to buy for the coming season. If you have lots of green beans left in March, take that as a clue that supply is exceeding demand. Put up less the coming year and fill that freezer space with something else.

Group like with like in your freezer and your inventory will be more organized, your searching simplified. One shelf or section can keep meats, another fruits, another cooked foods or main dishes. Devise an inventory form to help you keep track of where each category of food is. If you are a super-organized person, draw up a chart of food, date in, date out and use. Put the chart on a clipboard hung on the freezer door handle or nearby. Then note what goes in, out, how much and when. If you are more informal in record keeping, just tape a piece of paper on the front of the freezer, write down the quantity of what you put in, then make a mark when you take a package out. Do not forget to label each and every package clearly — in writing or symbols someone besides yourself can read! Legible labels, good packing or placement in the freezer make inventory and food selection easy.

As you use your frozen food, keep a running check on your methods and packaging. If you notice that a particular bag, container or sealing method is not doing the job, make a mental or actual note of it (why not right here in the margin?) and vow to change that procedure. If you do not like some of the food you have frozen, check into other methods.

Foods To Freeze

There are some foods that simply do not take to freezing. The list is short: lettuce, salad greens, radishes, green onions, custard or custard pies, hard-cooked egg white, meringues, egg white frostings, cakes with soft fillings, egg in the shell and raw potatoes.

Other than those, almost anything can be frozen. If you are not sure how you will like a food after it has been frozen, try a sample batch of just a few packages, bags or containers. Freeze for a couple of weeks, then taste-test. If you hate it, not much has been lost. If it works, you will feel confident to freeze it in quantity.

Certain varieties of fruits and vegetables freeze better than others, so check with your local agricultural extension office before you plant and/or buy. Knowledgeable produce people, either in the supermarket or at a stand, should be able to help, too.

Select the highest quality of food possible. Produce should be ripe or mature enough to eat right away, tender and as fresh as possible. For the very best flavor and texture, authorities suggest you transfer produce from the garden to the freezer within 2 hours! That means your equipment and kitchen have to be ready before you leave the house for the garden or market. If you cannot work within that time limit, keep produce chilled until you are ready to prepare it.

Produce to go in the freezer must be clean. Wash, scrub, rinse and drain it just as if you were going to eat it right away.

The Freezer

All our references to a freezer mean a separate appliance for freezing only. The ice-cube section of a refrigerator is good for very short term storage only, and "short term" means days, not weeks or months. The separate freezing compartment of a refrigerator can hold food for weeks; a side-by-side freezer section can hold food for a few months. But, for long-term storage at 0°F, a separate household freezer is your best bet.

The freezer size and type are up to you. A chest freezer usually costs less to buy, and to run, but an upright may fit into your home more easily. Most folks agree that it is easier to find food, and to get it out of an upright freezer. Check your space, your budget and your back to decide which type is best for you.

Figure on six cubic feet of freezer space per person in your family. Then, if you can manage it, buy a freezer bigger than that. Once you get used to having a freezer you will have no trouble filling it.

Keep track of your freezer's temperature with a refrigerator-freezer thermometer. Put it in front of the storage area, fairly high up in the load of food. Leave it overnight, without opening the freezer, before you check it for the first time. If thermometer reads above 0°F, adjust the freezer's temperature control to a lower setting. Leave freezer overnight and then check thermometer to see if you adjusted temperature correctly. When you have the temperature adjusted correctly, check the thermometer once a day. If your freezer has an automatic defrost, do not take a reading during the defrost cycle — it will not be an accurate reflection of normal freezer temperature.

Basic Equipment

Besides the freezer itself, the Basic Equipment for Freezing consists of various packaging materials, plus whatever kitchen implements you need to prepare the food for packaging.

Rigid freezer containers with tight-fitting lids for liquid foods: plastic freezer containers, shortening or coffee cans, freezer cans or jars with wide mouths
Bags: polyethylene, or heavy-duty plastic bags or boilable pouches
Ties for bags: string, rubber bands, pipe cleaners or twist ties
Shallow tray: cookie sheet or jelly roll pan for tray freezing
Freezer paper: moisture-vapor-proof, laminated or coated
Heavy-duty plastic wrap or heavy-duty aluminum foil
Freezer tape to seal wrapped foods
Labels (pressure sensitive, or use freezer tape)
Grease pencil or felt-tip marker

Packaging

The secret to successful freezer packaging is to get the air out and keep it out. Food exposed to dry frigid air dries out, devel-

Packaging Materials for Freezing. Use moisture-vapor proof wrappings to avoid freezer burn.

ops off flavors and can lose color and nutrients. Moisture on the surface of food freezes; the ice crystals, if exposed to air, evaporate and cause freezer burn. So, look for containers that are moisture and vapor proof.

Rigid freezer containers are best for foods that are liquid or do not have much shape of their own. These rigid containers can be plastic freezer containers with tight-fitting lids; canning or freezer jars with wide mouths and tight-fitting lids. Square or rectangular containers use freezer space more efficiently than round containers or those with flared sides or raised bottoms.

Bags made from polyethylene or heavy-duty plastic, or the new boilable pouches that can be heat sealed, can hold almost anything. Liquid foods are safest in plastic bags in protective cardboard boxes. Bags are not always easy to stack, but they are great for tray-frozen vegetables, bulky or odd-shaped items.

Freezer wrap is best for meat cuts, fish and game, casseroles and cakes. The wrap must be moisture and vapor proof. That means you should buy special laminated or coated freezer paper, heavy duty plastic wrap or heavy duty aluminum foil. Light-weight plastic wrap, butcher paper, waxed paper are not tough enough to protect food in the freezer. Freezer tape, not masking tape, must be used to seal wrapped food. Empty plastic-coated milk cartons, cottage cheese or ice cream containers are not airtight enough to be reused for freezer containers unless you bag and seal the food and then put it in a carton to protect the bag.

The best package size for you depends on your freezer and your family. Pack food in containers that will take care of your crew for one meal, remembering that giant containers take forever to thaw. Figure two servings to a pint container; three or four servings to a quart. It is quicker to thaw two pint containers than one larger container.

Head Space

Food expands as it freezes and you must allow room, or head space, for this expansion. Otherwise the lids will pop off, bags will burst and you will have wasted food, time and money. Foods that are dry, or have little or no added liquid, need ½-inch of space between the top of food and the top of the container for both pints and quarts. Food that is packed in liquid or is mostly liquid needs ½-inch head space for pints, 1 inch for quarts. Always use wide-mouthed containers. If you pack foods in containers with narrow mouths the food expands upward in the container even more, requiring ¾-inch head space for pints and 1½-inches for quarts. We suggest you stick to using wide-mouth containers. The recipes that follow give you head space needs for each particular food.

Sealing

How you seal food for the freezer is just as important as how you package it. Seal rigid containers by following the manufacturer's instructions (if there are any), or by snapping, screwing or fitting lid tightly on the container. If the lid does not seem tight, seal it with freezer tape.

Seal bags or boilable pouches with a heat sealing ap-

Step 1: Use enough wrap to go around the food 1-1/2 times.

pliance. Or seal bags by pressing out the air, then twisting the bag close to food. Fold the twisted section over and fasten it with a rubber band, pipe cleaner or twist tie. To get air out of an odd-shaped bag, lower the filled bag into a sink full of water and let the water press the air out. Twist the bag top, lift it out, double the twisted area backwards and fasten.

Drugstore Wrap

"Drugstore wrap" is the trick for properly sealing wrapped foods.

1. Tear off enough wrap to go around the food 1½ times.
2. Bring the ends together and fold them over, creasing along the fold.

3. Turn down, fold and crease and fold and crease until the folds are tight up against the food, pulling the wrap tightly against the food as you fold.
4. Press the paper down across food and out to the ends to press out the air. Fold the ends to points.
5. Fold the ends under about an inch or so, then fold the ends tightly against food.
6. Seal the ends to package with freezer tape.
7. Label with the date, food and weight or amount.

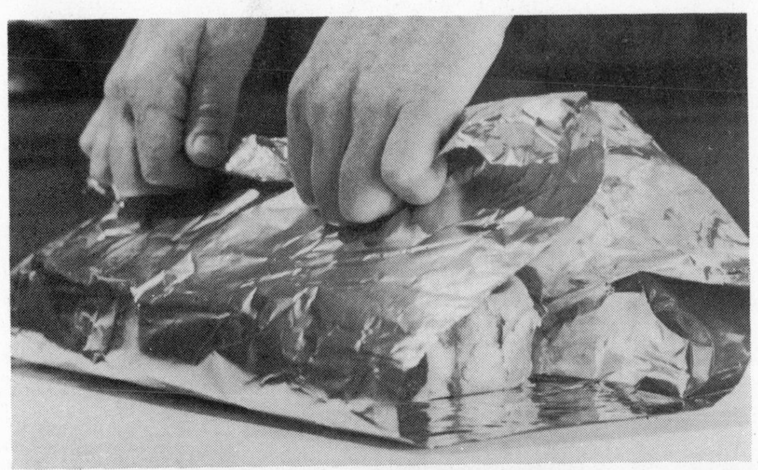

Step 2: Bring the ends together and fold them over, crease.

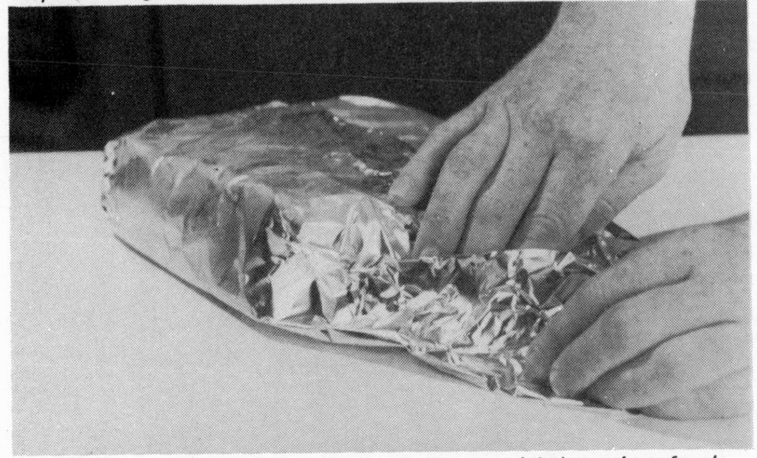

Step 3: Turn down fold and tuck side corners tightly against food.

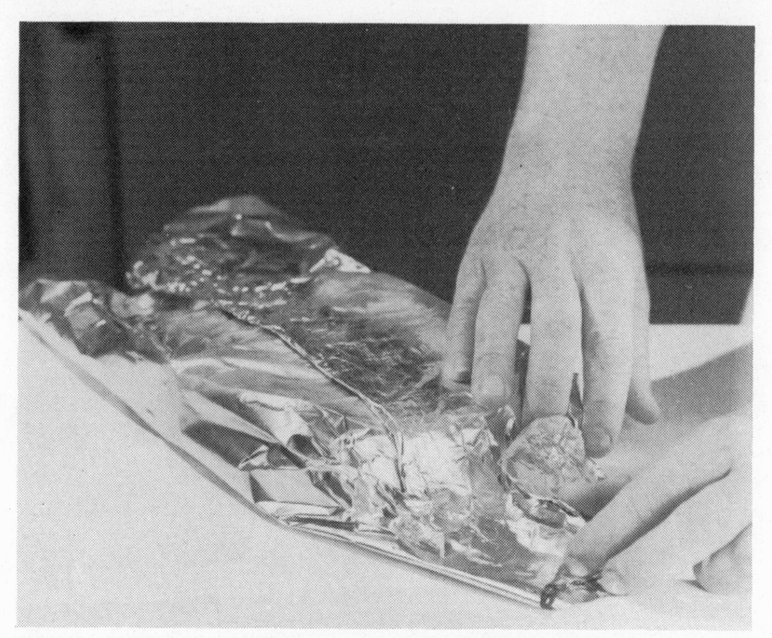

Step 5: After you press the air out, fold the ends to points.

Steps 6 and 7: Fold ends under about 1 inch, fold tightly against food then label package.

Butcher Wrap

The Butcher Wrap is another way to package some foods for the freezer. This method of wrapping works well for roasts and other chunky items.

1. Tear off enough freezer paper to go around the food twice.
2. Place the food across the corner and bring over the corner.
3. Roll the food and the paper toward the opposite corner until halfway down the paper.
4. Tuck the two side corners into the center of the package, pulling the wrap tightly and pushing the air out toward the open end.
5. Bring up the pointed end and seal well with freezer tape.
6. Label with the date, food and weight or amount.

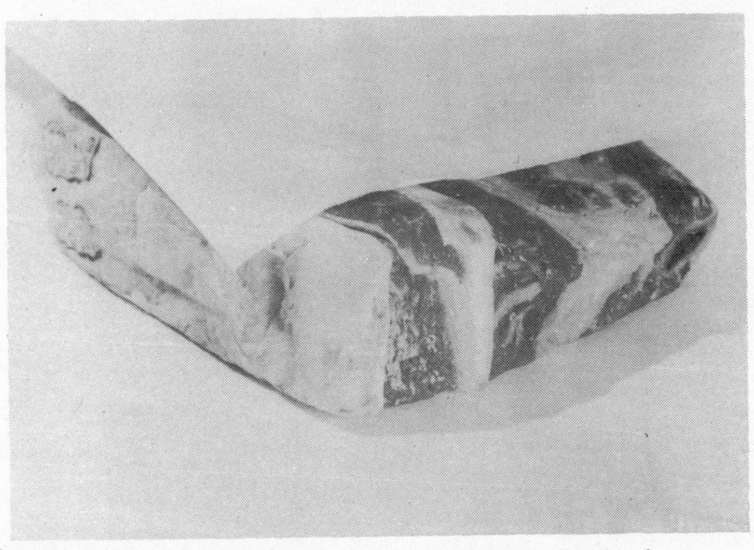

Step 1: Use enough freezer paper to go around the food twice.

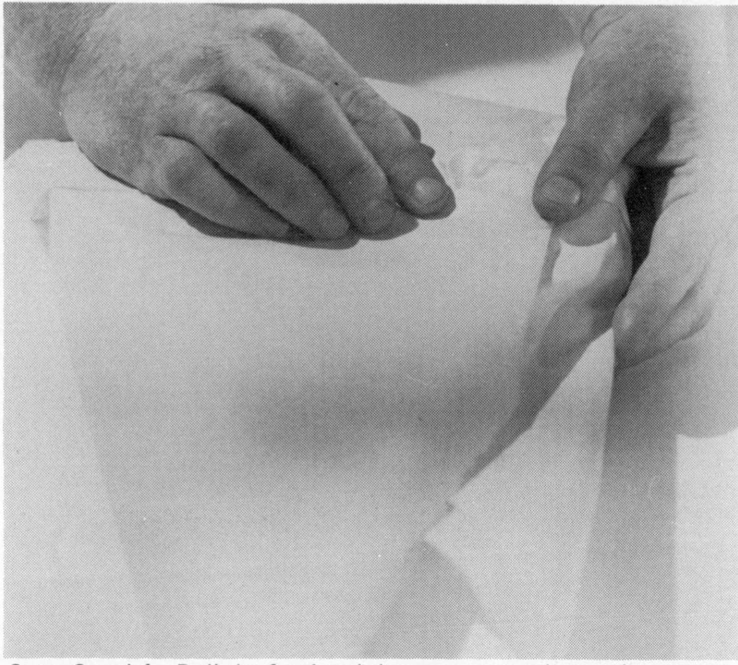

Steps 3 and 4: Roll the food and the paper toward opposite corner until halfway down the paper. Tuck in the inside corners.

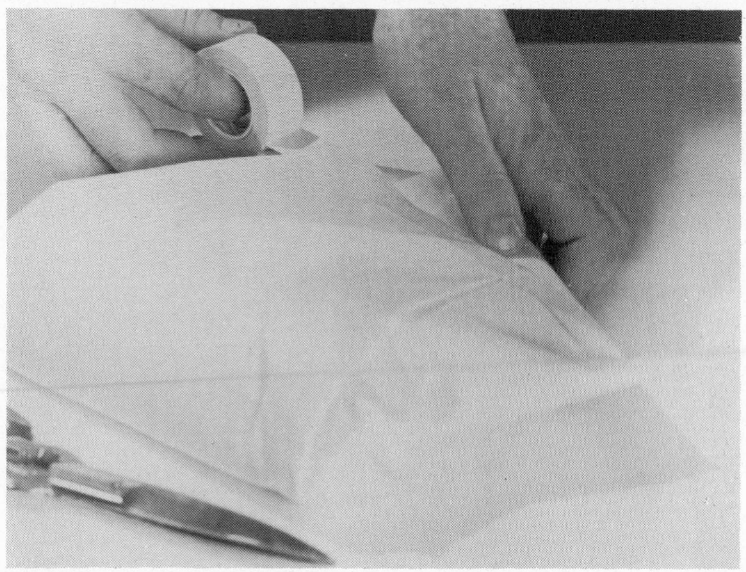

Step 5: Bring up pointed end and seal with freezer tape.

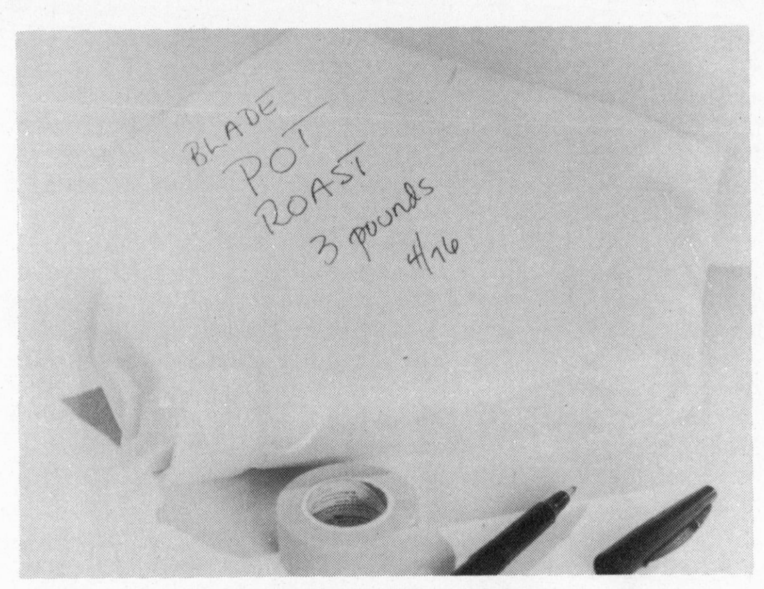

Step 6: Label with date, food, weight or amount.

Labeling

A good freezer label should tell what food is in the package and when it went into the freezer. Better yet, it should tell how the food was packed. For example, "Sugar Pack Strawberries June, 1976." Frozen main dishes, sauces packed in boilable pouches, and other more complex items call for a label with description, number of servings, perhaps even heating and thawing instructions.

Select labeling materials that will last. A grease pencil or felt tip marker may write directly on the container. Freezer tape makes a quick label, as do pressure sensitive labels from a stationery store. Try to print legibly and use standard, not personal, abbreviations.

Basic Rules

Freezing food is not just a matter of opening the door, tossing in packages of food and that's that. You have to approach the project scientifically to have good frozen food and to use energy wisely. Look back to the use and care book that came with your freezer to find the coldest sections of your freezer. You will probably find specific freezing directions, too. If you have an upright freezer the shelves are the coldest places; in a chest freezer, the coldness comes from the walls. Those are the spots where food to be frozen must go.

A few hours before you will be preparing food to freeze, set the freezer's control to its lowest setting. Arrange packages of cold food in a single layer on or near the coldest spots (or follow your freezer's directions), leaving a little space between each package for heat to escape. Shut the freezer and leave it alone for 24 hours.

After 24 hours the food should be frozen solid. Stack it up and move it away from the coldest part to another area in the freezer for storage. Then you can add another batch to be frozen.

Your freezer can only freeze a limited amount of food at a time, usually two to three pounds of food for each cubic foot of freezer space. Do not try to freeze any more than that or the food will freeze too slowly and quality will be lowered. The amount of food your freezer can freeze at once helps you determine how much food to prepare for the freezer on any particular day. If you have more food to be frozen than your freezer can take, either refrigerate packages for a day or so (not much longer) or cart it all to a locker to be frozen, then bring home for freezer storage. Always try to keep a cold spot free to quickly freeze additional food.

When you have finished freezing food in quantity, reset cold control to the setting that will maintain 0°F.

To Thaw

"There's nothing to thawing," you may say. But, how you thaw is important for food safety. Proper thawing will give you safe and tasty foods. Improper thawing might give you mushy meals and stomach trouble. The sections on foods that follow give you thawing specifics. Generally, always thaw food in the refrigerator. Never at room temperature (except for bread, baked goods and fruit). Meat, fish and poultry can be cooked without thawing — just add 1/3 to 1/2 more cooking time.

Emergency!

What if the power goes out or your freezer quits? DO NOT

OPEN THE FREEZER! First check the fuses and the plug; then read your instruction book, if you think something is wrong with the freezer. When you are certain there is a problem, or if the power is off, leave the freezer closed. A full freezer will keep food frozen for 15 to 20 hours and food will stay below 40°F for up to 48 hours.

If the freezer will be off for longer than that you have several alternatives. Quickly buy some dry ice and put it in. A 25-pound chunk of dry ice, carefully handled with gloves and placed on a piece of heavy cardboard on top of the packages of food, should hold a half-full freezer (10 cubic feet) for 2 to 3 days, or 3 to 4 days if the freezer is full. (Use 2½ pounds dry ice for each cubic foot.) If dry ice is unavailable, you can pack up the food and take it into a neighbor's freezer or a locker.

If food goes above 40°F (ordinary refrigerator temperature), check it over carefully and cook it completely immediately. You can refreeze the cooked food, but use it as soon as possible. If the food looks or smells suspicious and has been over 40°F, do not take any chances — toss it out. A freezer thermometer is an excellent guide to freezer safety. If you do not have a thermometer, feel the food and take a guess. Anything that still has ice crystals is safe to refreeze or use, if you are quick about it.

Freezing Fruit

Freezing captures the flavor and fragrance of the orchard, grove or patch better than any other preservation method. You prepare food for the freezer just as if you were preparing it to eat right away: wash, peel, pit, slice or cut and sweeten. Although fruits can be frozen unsweetened (and be sure to try this if you have calorie-watchers or special dieters in your family) most have better flavor, color and texture if packed with sugar or syrup.

Because many fruits darken when cut and exposed to air, you should give them a simple but special treatment with ascorbic acid or commercial color-keeper (Fruit Fresh) to preserve color.

Basic Equipment

Besides the Basic Equipment for Freezing, you will need the following equipment to prepare fruit for freezing.

Colander or strainer
Paring and chopping knives
Measuring cups for fruit, sugar or syrup
Masher
Chopper or grinder
Mixing spoons and measuring spoons
Large mixing bowl or pan to put the fruit in as you prepare it. This container must be aluminum, unchipped enamel, glass, stainless steel or crockery. Do not use galvanized iron, tin or chipped enamel ware.

Basic Ingredients

In addition to the fruit — fresh, perfect, ripe and ready-to-eat — sugar or a syrup are the only ingredients called for in freezing fruit. Only white granulated sugar is used for sugar-packed fruit; but, for syrup-packed fruit, you can substitute light corn syrup or mild-flavored honey for ¼ of the total amount of the sugar.

Packing Fruit

Fruits are packed for the freezer in four ways: sugar, syrup, unsweetened, or tray pack.

Sugar pack is the best method for juicy fruits and for fruits that do not darken. Use this pack for fruits you are going to use later in cooking. Always use white granulated sugar. The

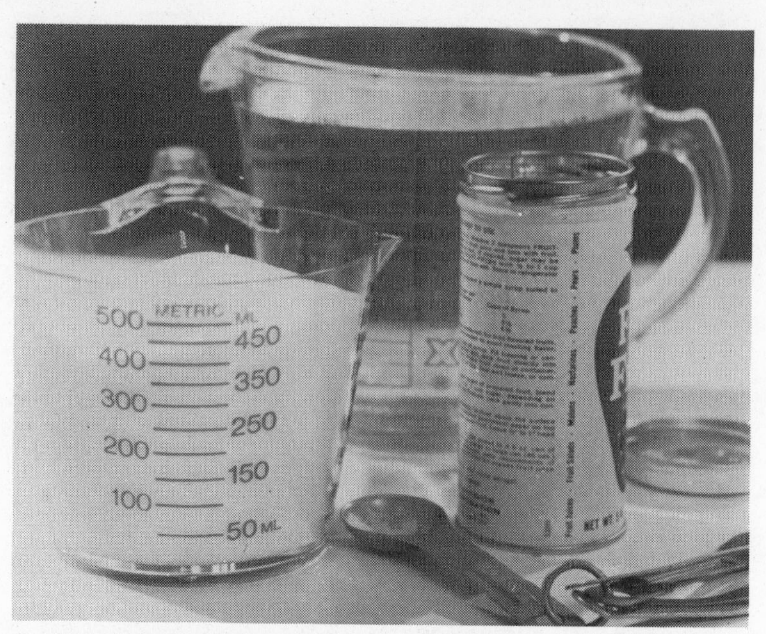

Fruits have better flavor and shape if frozen in sugar or syrup-ascorbic acid mix.

Sugar Pack for Fruit. Sprinkle sugar over prepared fruit, mix gently until sugar dissolves.

Layered Sugar Pack. For juicy fruits, alternate layers of prepared fruit and sugar in rigid container. Shake to distribute sugar.

following recipes give specific amounts for each fruit. Sprinkle sugar over the prepared fruit, mix gently until sugar dissolves and juices start to flow. Then spoon the fruit and juices into wide-mouth pint or quart containers, filling to within ½ inch of the tops of pints, 1 inch of the tops of quarts. Crumple a small piece of plastic wrap, freezer or waxed paper and put it on top of the fruit to hold the fruit down in juice. Wipe the sealing edge of the container with a clean damp cloth, put the lid in place and seal. Label and freeze.

A step-saving variation on the sugar pack lets you alternate layers of prepared fruit and sugar in containers. Check each recipe for the exact amount of sugar; it is usually about ¼ cup for each pint. Shake the containers to distribute sugar,

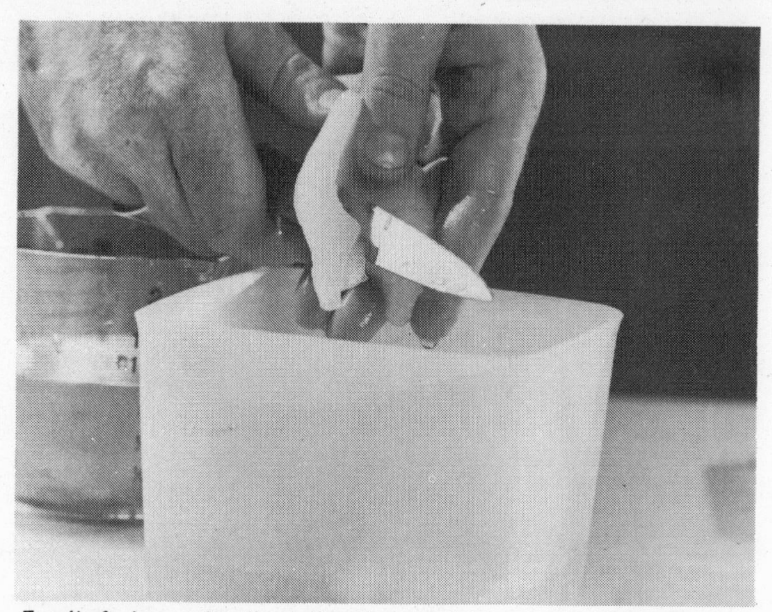

Fragile fruits can be sliced directly into rigid container filled with cold syrup.

Keep fruit submerged in syrup with a crumpled piece of freezer paper, plastic wrap, or waxed paper.

then seal and freeze as directed above.

Syrup pack is the choice for fruits to be served as desserts or salads or for fruits that discolor. Prepare syrup in advance, following the chart. Since the syrup must be cold, you can mix it up the day or night before. If you wish, substitute ½ of the total amount of sugar with light corn syrup or light, mild-flavored honey. Pack fruit gently but firmly into containers, pressing out the air. Pour syrup over fruit and fill to within ½ inch of top of wide-mouth container for pints, 1 inch for quarts. When working with peaches or other fragile fruits you can pour ½ cup of cold syrup into each container, then slice the fruit right into the syrup. Gently press fruit down into the container and add enough syrup to cover it. Crumple a small

Sealing to keep air out is important. Containers must have tight-fitting lids.

piece of plastic wrap, waxed paper or freezer paper and put it on top of the fruit to hold the fruit down in the syrup. Wipe the sealing edge of the container with a clean damp cloth, put the lid in place and seal. Label and freeze.

To make syrup, combine sugar and water (or sugar, corn syrup or honey) in the proportions given in the chart. Mix until the sugar dissolves. Chill until ready to use.

Type syrup	Sugar cups	Water cups	=	Syrup cups
30%	2	4		5
35%	2½	4		5⅓
40%	3	4		5½
50%	4¾	4		6½
60%	7	4		7¾
65%	8¾	4		8⅔

Fruits can be frozen unsweetened. Here, melon balls are sealed in pouches.

Count on ½ to ⅔ cup of syrup for each pint of fruit, thus 1 quart of syrup will take care of 6 to 8 pints. Most recipes call for a 40% syrup, so the 5½ cups of syrup above should be enough for 8 to 11 pints of fruit. Refrigerate any leftover syrup to use another time — for freezing or for sweetening beverages or other foods.

Unsweetened pack can be used for whole, sliced or crushed fruits. Pack whole or sliced fruits into containers, seal and freeze. If fruits darken, pack them into containers, then cover with water mixed with ascorbic acid (¼ to ½ teaspoon to a quart of water), leaving ½-inch head space for pints, 1 inch for quarts. In place of water you can use fruit puree or juice. Pack crushed fruits into containers, filling to within ½-inch of the top of pints, 1 inch of the top of quarts. Seal and freeze.

Tray pack is great for small whole fruits. Wash and drain

Tray Pack. Spread fruit on tray, freeze solid, then transfer to containers for freezer storage.

the fruit, hull or stem, if necessary, then spread a single layer on a cookie sheet or shallow tray and freeze solid. Working quickly, pack the fruit in bags or containers, seal and return to the freezer to store. When ready to use, just pour out the amount you need.

To Prevent Darkening

Peaches, apples, nectarines, apricots and a few other fruits will turn brown when exposed to air, unless treated with ascorbic or citric acid, lemon juice or a commercial product, such as Fruit Fresh (follow label directions).

Ascorbic acid is probably the cheapest treatment. Citric acid or lemon juice may overpower the delicate fruit flavor and may also make fruit too tart for many tastes. Ascorbic acid is a standard drugstore item these days. The crystalline or powdered form is better for freezing use than tablets; and our recipes call for crystalline or powdered ascorbic acid. The tablets are hard to dissolve — you have to crush them first — and the filler used to hold them together can cloud the syrup. If you can only get tablets, follow the substitution chart.

To use ascorbic acid: dissolve the crystals, powder or crushed tablets in ¼ cup cold water just before adding to the fruit.

For sugar pack, sprinkle dissolved ascorbic acid over the fruit just before adding the sugar. Use ¼ teaspoon ascorbic acid dissolved in ¼ cup of cold water for each quart of prepared fruit.

For syrup pack, gently stir dissolved ascorbic acid into the cold syrup just before pouring it over the fruits or into the containers.

For unsweetened pack, dissolve ¼ to ½ teaspoon ascorbic acid in 1 quart cold water and pour it over the fruit to cover.

Crystalline or powdered ascorbic acid, teaspoons	=	ascorbic acid tablets, milligrams
⅛		375
¼		750
½		1500
¾		2250
1		3000

Basic Steps

1. Check the freezer space to see how much food your freezer can handle. Remember that you should add no more than two to three pounds of food to be frozen for each cubic foot of freezer space. The amount of space you have available will determine how much fruit you pick or buy and prepare. If you have more fruit than space, check with a locker plant to see if they have space.
2. If using a syrup pack, prepare the syrup ahead so it can be chilled.
3. Choose perfect, ripe, ready-to-eat, fresh fruit. Keep the fruit chilled until ready to prepare.
4. Set out the containers, lids and other equipment. Be sure all the equipment is clean.
5. Work with small amounts of fruit at a time, especially those fruits that darken. Wash and prepare enough fruit for several containers (the recipes tell you how much fresh equals how many pints frozen), pack and freeze, then start over again.
6. Wash the fruit well in plenty of cold water, handling gently to prevent bruising, but scrubbing to remove dirt, if necessary. Always lift the fruit out of water, so dirt stays in the water and is not redeposited on the fruit. Use a colander or strainer to hold small fruits while they drain. Drain large fruit on the sink drainboard or on paper towels. Do not let fruit stand in water any longer than necessary or it will lose nutrients.
7. Sort the fruit by size, shape and ripeness and keep similar fruits together for preparation.
8. Prepare the fruit as the recipe directs.
9. Pack the fruit into containers using the sugar pack, syrup pack, unsweetened pack, or tray pack, as recipe directs. Leave the head space so that when fruit expands during freezing it will not push the lid off and break the seal. The recipes will tell you how much head space to leave.
10. Wipe off the edges and rims of the containers with a clean damp cloth, so the sealing areas are clean.
11. Seal the rigid plastic containers by closing the lid firmly or as the manufacturer directs. Seal the bags by pushing out all the air possible; twist the bag just above the fruit, then loop the twisted portion backward and fasten it with a rubber band, string or twist tie.
12. Label the fruit clearly with the name of the fruit, type of

pack, date put in the freezer or day by which to use it. Most fruits will maintain good quality for 8 to 12 months at 0°F. Citrus fruits and juices are best used within 4 to 6 months.

To Thaw Frozen Fruit

Do not uncover the containers of fruit until fully thawed. Refrigerator thawing takes about five to seven hours, room temperature about two to four hours. (Fruit is the only food recommended to thaw at room temperature.) An unopened container can be placed in a bowl of cool water and it will thaw in about one hour. Fruit packed in boilable pouches will thaw in 15 to 20 minutes. Do not overthaw. Frozen fruit is best served while still frosty with some ice crystals.

Recipes

APPLES

Pack apples in syrup for desserts, in sugar for pie-making.

1¼-1½ pounds = 1 pint, frozen
1 box (44 pounds) = 32-40 pints, frozen
1 bushel (48 pounds) = 32-40 pints, frozen

1. Choose crisp, firm and flavorful apples with no bad or bruised spots.
2. Wash, peel and core.
3. Syrup Pack: Gently stir ½ teaspoon ascorbic acid into each quart of 40% syrup. Pour ½ cup syrup into each freezer container and slice the apples into the container. Gently push the apple slices down to fill the containers and press out any air. Pour in additional syrup, enough to cover the slices, leaving ½-inch head space for pints, 1 inch for quarts. Crumple a small piece of plastic wrap, waxed paper or freezer paper and put it on top of the apples in each container to hold the apples under the syrup. Seal and freeze.

Sugar Pack: Stir 2 tablespoons salt into a gallon of water and slice the apples into the salt water to prevent darkening. When you have sliced enough to fill several containers, drain the apples. (They should not stand in salt water for more than 15 minutes.) Measure the apple slices into a large bowl or shallow pan and sprinkle with ½ to 1 cup sugar for each quart of slices. Mix gently until the sugar dissolves. Pack the apples gently but firmly into containers, leaving head space and topping with a piece of paper as for the syrup pack. Seal and freeze.
Unsweetened Pack: Slice the apples into salt water, as for the Sugar Pack, then pack the slices into containers, leaving ½ inch of head space for pints and quarts. Seal and freeze. As an extra precaution against darkening, steam-blanch the apple slices before packing them unsweetened. If you do not have a steamer, put a rack in the bottom of any wide-bottomed pan with a tight-fitting cover. Pour in enough water to cover the bottom but not touch the rack. Heat the water to boiling, then spread apple slices in single layer on the rack. (Put a cheesecloth over the rack if the slots are wide.) Cover and steam 2 minutes. Drain, dip the fruit in cold water to cool; drain again and pack following the unsweetened pack method.

APPLESAUCE

For quick thawing, seal the sauce in boilable pouches — 15 minutes in a bowl of cool water and it is table-ready. Add spices of your choice after thawing. You will need a food mill or strainer, to make smooth applesauce.

2 pounds = about 1 quart, frozen

1. Choose ripe, flavorful apples.
2. Wash, core and slice the apples. Peel, if you wish.
3. Cook each quart of sliced apples with 1/3 cup water and ½ to ¾ cup sugar just until tender.
4. Cool the apples completely. Put them through a food mill or strainer, if you wish.
5. Pack the sauce into containers, leaving 1 inch of head space for pints and quarts.
6. Seal and freeze.

APRICOTS

If you are freezing apricots in a sugar pack for pie-making you do not need to peel them.

2/3 to 4/5 pound = 1 pint, frozen
1 crate (22 pounds) = 28 to 33 pints, frozen
1 bushel (48 pounds) = 60 to 72 pints, frozen

1. Choose firm, ripe, evenly colored apricots.
2. Wash, halve and pit. Peel and slice, if you wish. If you do not peel the apricots you may wish to dip them in boiling water 30 seconds, just to tenderize skins. Then dip them in cold water and drain.
3. Syrup Pack: Gently stir ¾ teaspoon ascorbic acid into each quart of 40% syrup. Pour ½ cup syrup into each freezer container. Fill with apricots, packing gently but firmly. Add more syrup, enough to cover, leaving ½ inch of head space for pints, 1 inch for quarts. Crumple a small piece of plastic wrap, waxed paper or freezer paper and put it on the apricots in each container to hold them under syrup. Seal and freeze.
   Sugar Pack: Dissolve ¼ teaspoon ascorbic acid in ¼ cup cold water and sprinkle over each quart of prepared apricots. Then sprinkle each quart apricots with ½ to 2/3 cup sugar. Mix gently until the sugar dissolves. Pack gently but firmly into containers, pressing down so the apricots are covered with juice. Leave head space and put paper on top as in the syrup pack. Seal and freeze.

APRICOT PUREE

This puree is an elegant sauce for ice cream, puddings or cakes. You can also use it in chiffon pies, Bavarian creams, or salads.

1. Choose fully ripe fruit.
2. Wash, halve and pit. Measure the halves.

3. Heat with just a few tablespoons water over medium heat until just simmering.
4. Press through sieve or food mill or puree in blender.
5. Dissolve ¼ teaspoon ascorbic acid in ¼ cup cold water for each quart of halves; stir in the puree.
6. Stir in 1 cup sugar for each quart of halves.
7. Pack into containers, leaving ½ inch of head space for pints, 1 inch for quarts.
8. Seal and freeze.

AVOCADOS

Yes, you can freeze avocados. Freezing is a handy way to tuck this elegant fruit aside when your tree is full or prices are low. Use thawed frozen puree for dips and salads.

1. Choose soft, ripe but not mushy avocados.
2. Peel, halve and seed.
3. Puree by pressing through strainer or food mill or in blender. Measure puree.
4. Stir in ⅛ teaspoon ascorbic acid for each quart puree.
5. Pack in pint containers, leaving ½ inch of head space.
6. Seal and freeze.

BANANAS

Freezing is a great way to save bananas that have ripened too much for out-of-hand eating. Package in the amounts you need for your favorite banana bread or cake recipe, then bake when you want to, not when you have to!

1. Choose soft-ripe bananas.
2. Peel and mash with 2 tablespoons lemon juice or ½ teaspoon ascorbic acid for each 1 to 2 cups mashed bananas.
3. Measure amounts used in your favorite recipes.
4. Pack into containers, leaving ½ inch of head space.
5. Seal and freeze.

BERRIES (blackberries, blueberries, boysenberries, dewberries, elderberries, gooseberries, huckleberries, loganberries and youngberries)

Blueberries are great for unsweetened or tray freezing. Other berries can be packed by any method, but syrup is best.

1½ pints = 1 pint frozen
1 crate (24 quarts) = 32 to 36 pints frozen

1. Choose ripe, plump berries.
2. Pick over and discard any imperfect berries. Stem, if necessary.
3. Wash in plenty of cold water. Lift out of water and drain in colander or on paper towels.
   Sugar Pack: Sprinkle ¾ cup sugar over each quart of berries. Mix gently until the sugar dissolves. Pack gently but firmly into containers, leaving ½ inch of head space for pints, 1 inch for quarts. Crumple small piece of plastic wrap, waxed or freezer paper and put on top of berries in each container. Seal and freeze.
   Syrup Pack: Pack berries gently but firmly into the container and cover with cold 40% syrup (use 50% if the berries are tart), leaving a head space and topping with paper as in the sugar pack. Seal and freeze.
   Unsweetened Pack: Pack the berries in containers, leaving head space as in the sugar pack. Seal and freeze.
   Tray Pack: Spread the berries in a single layer on a cookie sheet, jelly roll pan or shallow tray. Freeze solid. When frozen, quickly pack the containers or bags and store in the freezer.

CHERRIES, TART RED

Sugar pack cherries for pies.

1¼ to 1½ pounds = 1 pint frozen
1 bushel (56 pounds) = 36 to 44 pints frozen

1. Choose tree-ripened cherries that are bright red.
2. Stem, discard imperfect cherries and wash thoroughly. Drain and chill before pitting. Pit with cherry pitter or clean hairpin.
   Sugar Pack: Sprinkle each quart of cherries with ¾ to 1 cup sugar and mix gently until the sugar dissolves. Pack gently but firmly into containers, leaving ½ inch of head space for pints, 1 inch for quarts. Crumple a small piece of plastic wrap, waxed paper or freezer paper and put it on top of the cherries in each container. Seal and freeze.
   Syrup Pack: Pack the cherries gently but firmly into the container and cover with cold 60% or 65% syrup, leaving head space and adding paper as in the sugar pack. Seal and freeze.

CHERRIES, SWEET

The red, rather than dark, sweet cherries are best for freezing.

Fresh to frozen yields are same as for Tart Red Cherries.

1. Choose tree-ripened, sweet, beautifully colored cherries.
2. Stem, discard imperfect cherries and wash thoroughly. Drain and chill before pitting.
3. Pit with a cherry pitter or clean hairpin, if desired. Pitting is not necessary, but the pit does add an almond-like flavor to the fruit.
4. Syrup Pack: Pack cherries gently but firmly into containers. Gently stir ½ teaspoon ascorbic acid into each quart of cold 40% syrup and pour it over the cherries, leaving ½ inch of head space for pints, 1 inch for quarts. Crumple a small piece of plastic wrap, waxed paper or freezer paper and put it on top of the cherries in each container. Seal and freeze.

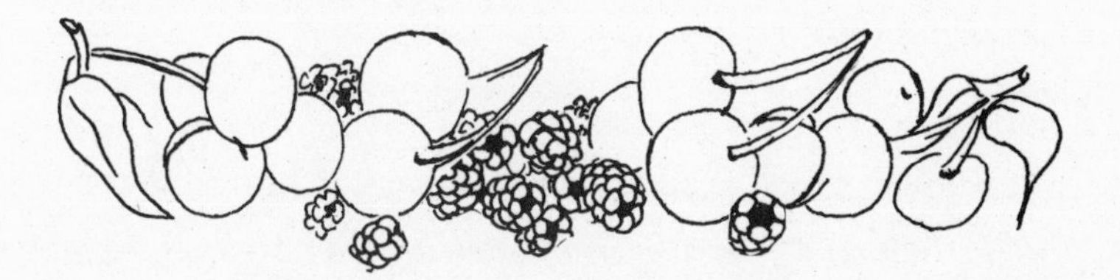

COCONUT

Keep coconut on hand for desserts, curries, cookies and cakes.

1. Break a fresh coconut, reserving the milk if possible.
2. Shred the coconut meat or put it through a food grinder or chopper.
3. Pack into containers, leaving ½ inch of head space for pints and quarts. If you saved the milk, pour it over the coconut in the containers, if you wish, leaving ½ inch of head space for pints, 1 inch for quarts. Seal and freeze.

CRANBERRIES

These bouncy red berries get the "Easiest-Ever Freezing Award." Just buy them in polyethylene bags at the supermarket, then hurry home and pop the bags into the freezer. Wash and sort when you are ready to use them.

DATES

Packages of dates from the store can go right in the freezer and wait there till you need them.

Wash fresh dates, slit and remove the pits, then pack them into containers, seal and freeze.

FIGS

Use a syrup pack to save figs for desserts and fruit cups.

1. Choose soft-ripe, tree-ripened fruit. Sample one to be sure

figs have not soured in the center.
2. Wash, stem and peel, if you wish. Slice or leave the figs whole.
3. Syrup Pack: Pack figs into containers. Gently stir ¾ teaspoon ascorbic acid into each quart of 35% syrup and pour it over the figs, leaving ½ inch of head space for pints, 1 inch for quarts. Crumple a small piece of plastic wrap, waxed paper, or freezer paper and put it on top of the figs in each container.
   Unsweetened Pack: Pack figs into containers. Seal and freeze. If you wish, dissolve ¾ teaspoon ascorbic acid in each quart of water and pour over the figs in containers, leaving ½ inch of head space for pints, 1 inch for quarts. Seal and freeze.

FRUIT COCKTAIL

Put together your own combination of fruits and you will have an appetizer or dessert to rival the fanciest of restaurants.

1. Use sliced or chunked peaches, pears or apricots; melon balls or chunks; orange or grapefruit sections; seedless green grapes; dark sweet cherries; pineapple chunks or cubes; blueberries; plums or prune-plums.
2. Syrup Pack: Pack the fruit into containers. Cover with cold 30% or 40% syrup (depending on the sweetness of the fruits), leaving ½ inch of head space for pints, 1 inch for quarts. Crumple a small piece of plastic wrap, waxed paper or freezer paper and put it on top of the fruit in each container. Seal and freeze.

GRAPES

Freeze grapes in syrup unless you are going to use the grapes for juice or jelly, then pack them unsweetened.

1. Choose flavorful, firm but ripe grapes with tender skins.
2. Wash and stem. Halve and seed, if necessary. Leave seedless grapes whole.
3. Syrup Pack: Pack grapes into a container and cover with a cold 40% syrup, leaving ½-inch of head space for pints, 1 inch for quarts. Crumple a small piece of plastic wrap, waxed paper or freezer paper and put it on top of the grapes in each container. Seal and freeze.

   Unsweetened Pack: Pack the grapes into containers, leaving ½-inch of head space for pints and quarts. Seal and freeze.

   Tray Pack: Spread a single layer of grapes on a cookie sheet, jelly roll pan or shallow tray. Freeze solid. When frozen, quickly pack the grapes in containers or bags and store in the freezer.

GRAPEFRUIT AND ORANGES

Frosty cold grapefruit or orange sections can perk up breakfasts or can be used in salads or as meat accompaniments.

1. Choose firm, heavy, tree-ripened fruit without any soft spots.
2. Wash, then peel, removing all white membrane.
3. Section and remove seeds. You can slice oranges instead of sectioning, if you wish.
4. Syrup Pack: Gently stir ½ teaspoon ascorbic acid into each quart cold 40% syrup. Pour ½ cup syrup into each container. Pack in slices or sections, then add more syrup, enough to cover, leaving ½-inch of head space for pints, 1 inch for quarts. Crumple a small piece of plastic wrap, waxed paper or freezer paper and put it on top of the fruit in each container. Seal and freeze.

MELON (cantaloupe, crenshaw, honeydew, Persian, watermelon)

Frozen melons are best served when still frosty, otherwise the melon pieces get limp.

1 to 1¼ pounds = 1 pint, frozen
28 pounds = 22 pints, frozen

1. Choose ripe melons with firm but tender flesh and beautiful color.
2. Halve, seed and peel, then cut the melon into chunks, slices or balls.
3. Syrup Pack: Pack melon pieces into containers and cover with cold 30% syrup, leaving ½-inch of head space for pints, 1 inch for quarts. Crumple a small piece of plastic wrap, waxed paper or freezer paper and put it on top of the fruit in each container. Seal and freeze.

PEARS

Some varieties do not freeze well, so check to see which type you have before freezing. Plan on preparing more syrup than for other fruits, since you need it for pre-cooking the pears.

1 to 1½ pounds = 1 pint, frozen
1 western box (46 pounds) = 37 to 46 pints, frozen
1 bushel (50 pounds) = 40 to 50 pints, frozen

1. Wash the pears, then peel, halve or quarter and core.
2. Heat 1 quart 40% syrup to boiling for approximately every 5 pounds pears. Add the pears and cook 2 minutes.
3. Drain, reserving the syrup.
4. Chill the pears and syrup.
5. Syrup Pack: Pack the cold pears into containers. Gently stir ¾ teaspoon ascorbic acid into each quart cold 40% syrup, pour it over the pears, leaving ½-inch of head space for pints, 1 inch for quarts. Crumple a small piece of plastic wrap, waxed paper or freezer paper and put it on top of the fruit in each container. Seal and freeze.

PEACHES AND NECTARINES

1 to 1½ pounds = 1 pint, frozen
1 lug (20 pounds) = 13 to 20 pints, frozen
1 bushel (48 pounds) = 32 to 48 pints, frozen

Prepare and pack as for apricots. Nectarines do not need to be peeled, but peaches should be. Gently rub the surface of a peach with the back or dull side of a paring knife to loosen the skin, then peel. This gives a better frozen peach than dipping it in boiling water to loosen the skin.

For a delightful difference, use reconstituted frozen lemonade concentrate or orange juice to cover the peaches or nectarines instead of syrup.

PERSIMMONS

Puree persimmons to freeze, then use them in puddings, cakes or muffins.

1. Choose ripe, soft, orange persimmons.
2. Wash, peel and cut them into sections.
3. Press through a sieve or food mill or puree in a blender.
4. Measure the puree.
5. Stir ⅛ teaspoon ascorbic acid into each quart puree.
6. Unsweetened Pack: Pack puree into containers, leaving ½-inch of head space for pints and 1 inch for quarts. Seal and freeze.
   Sugar Pack: Stir 1 cup sugar into each quart puree. Pack into containers, leaving head space as in the unsweetened pack. Seal and freeze.

PINEAPPLE

For more flavor, use pineapple juice for part of the water when you make syrup for the Syrup Pack.

5 pounds fresh = 4 pints frozen

1. Cut off the top of the pineapple, then pare. Cut out the core and eyes.
2. Slice, chunk, cube or cut it into sticks or spears.
3. Unsweetened Pack: Pack into containers, leaving ½-inch of head space for pints and quarts. Seal and freeze.
   Syrup Pack: Pack tightly into containers and cover with cold 30% syrup, leaving ½-inch of head space for pints, 1 inch for quarts. Crumple a small piece of plastic wrap, waxed paper or freezer paper and put it on top of the fruit in each container. Seal and freeze.

PLUMS AND PRUNE-PLUMS

Pack plums and prunes unsweetened to make into jams or pies, or pack them in syrup for salads or desserts.

1 to 1½ pounds = 1 pint, frozen
1 crate (20 pounds) = 13 to 20 pints, frozen
1 bushel (56 pounds) = 38 to 56 pints, frozen

1. Choose tree-ripened fruit that is firm and beautifully colored.
2. Wash the fruit well. Leave it whole or halve or quarter and pit.
3. Unsweetened Pack: Pack whole fruit into containers, leaving ½-inch head space for pints and quarts. Seal and freeze.
   Syrup Pack: Pack halved or quartered fruit in containers. Gently stir ½ teaspoon ascorbic acid into each quart cold 40% or 50% syrup (depending on the sweetness of the plums). Pour it over plums in the containers, leaving ½-inch of head space for pints, 1 inch for quarts. Crumple a small piece of plastic wrap, waxed paper or freezer paper and put on top of the fruit in each container.
   Tray Pack: Spread whole or halved plums in single layer on a cookie sheet, jelly roll pan or shallow tray. Freeze solid. When frozen, quickly pack the plums into containers or bags and store in the freezer.
   Puree: Prepare as for Apricot Puree. If plums are soft-ripened you can omit heating before pureeing.

RASPBERRIES

Pick your method — all work equally well for raspberries, or follow the recipe for Raspberry Puree.

1 pint = 1 pint, frozen
1 crate (24 pints) = 24 pints, frozen

1. Choose ripe, juicy and fragrant berries.
2. Wash them gently in cold water and drain well.
3. Sugar Pack: Sprinkle ¾ cup sugar over each quart berries and mix gently so you do not crush them. Pack gently into containers, leaving ½-inch of head space for pints, 1 inch for quarts. Seal and freeze.
   Syrup Pack: Gently pack the berries into containers and cover with cold 45% syrup, leaving ½-inch of head space for pints and 1 inch for quarts. Crumple a small piece of plastic wrap, waxed paper or freezer paper and put it on top of the fruit in each container. Seal and freeze.
   Unsweetened Pack: Pack berries gently in containers, leaving ½-inch of head space for pints and quarts. Seal and freeze.
   Tray Pack: Spread the berries in single layer on cookie sheet, jelly roll pan or shallow tray. Freeze solid. When frozen, quickly pack the berries into containers or bags and store in freezer.

RASPBERRY PUREE

Just like fresh raspberries, this puree is superb for Melba or Cardinal sauce, sundaes, chiffon pies, puddings or for making jelly or jam.

1. Choose berries and prepare as for Raspberries, then puree through a food mill or sieve or in blender.
2. Measure the puree.
3. Stir in ¾ to 1 cup sugar for each quart of puree. Mix until the sugar dissolves.
4. Pack into containers, leaving head space as for sugar packed raspberries. Seal and freeze.

RHUBARB

This spring fruit will probably be the garden's first contribution to your freezer.

2/3 to 1 pound = 1 pint, frozen
15 pounds = 15 to 22 pints, frozen

1. Select colorful, firm but tender stalks with few fibers.
2. Wash, trim off leaves and woody ends.
3. Cut in 1- or 2-inch lengths.
4. Blanch 1 minute in boiling water, then cool quickly in cold water to save the color and flavor.
5. Unsweetened Pack: Pack rhubarb pieces tightly in containers, leaving ½-inch head space for pints and quarts. Seal and freeze.

   Syrup Pack: Pack rhubarb pieces tightly in containers. Cover with cold 40% syrup, leaving ½-inch of head space for pints, 1 inch for quarts. Crumple a small piece of plastic wrap, waxed paper or freezer paper and put it on top of the rhubarb in each container. Seal and freeze.

STRAWBERRIES

Strawberries are the all-time favorite fruit for freezing. Unsweetened berries do not have quite the quality of those frozen in a syrup or sugar pack, but they are still good.

2/3 quart = 1 pint, frozen
1 crate (24 quarts) = 38 pints, frozen

1. Choose firm, ripe, red berries. Slightly tart berries are alright.
2. Wash gently in plenty of cold water, a quart at a time.
3. Lift the berries out of the water and drain in a colander or on paper towels.
4. Hull, removing only the top, leaving as much fruit as possible.
5. Slice or crush large berries. Slice smaller berries, if you wish.
6. Sugar Pack: Slice berries into a large bowl or pan. Sprinkle on ¾ cup sugar for each quart berries. Mix gently until the juices begin to flow. Pack into containers, leaving ½-inch head space for pints, 1 inch for quarts. Crumple a small piece of plastic wrap, waxed paper or freezer paper and put it on top of the berries in each container. Seal and freeze.
   Syrup Pack: Pack whole or sliced berries gently but firmly into containers. Cover with a cold 50% syrup, leaving head space and topping each container with paper as in Sugar Pack. Seal and freeze.
   Unsweetened Pack: Pack whole berries into containers, leaving ½-inch head space for pints and quarts. Seal and freeze.
   Tray Pack: Spread whole berries in a single layer on a cookie sheet, jelly roll pan or shallow tray. Freeze solid. When frozen, quickly pack the berries into containers or bags and store in the freezer.

STRAWBERRY PUREE

A super sundae sauce, this puree could also be the basic ingredient for Strawberry Chiffon Pie or Bavarian Cream, or for making mid-winter jam or jelly.

1. Choose berries as for freezing, wash and hull.
2. Puree in blender or press through a sieve or food mill.
3. Measure the puree.
4. Stir in $^{2}/_{3}$ cup sugar for each quart puree. Stir until the sugar dissolves.
5. Pack into containers, leaving head space as for sugar packed strawberries.
6. Seal and freeze.

FRUIT JUICES (non-citrus)

Put away juices for punches, for drinking, or for making jelly later.

1. Choose ripe, soft fruits (cherries, berries, plums or prunes, grapes).
2. Heat the fruit over medium heat until the juices begin to separate from the fruit.
3. Strain through a jelly bag or several thicknesses of cheesecloth.
4. Sweeten to taste with sugar, about 1 cup sugar to each gallon juice.
5. Pour into containers (glass is best), seal and freeze. Or, pour into ice cube trays and freeze. When frozen, pack the juice-cubes in plastic bags. Seal and store in the freezer.

FROZEN FRUIT PIE FILLING

While preparing fruit for the freezer, fix some for pies. It is easy! The recipe makes one 9-inch pie.

- 1 quart blueberries, pitted cherries, sliced strawberries or peaches, apples or pears
- ¼ to ½ teaspoon ascorbic acid
- 1 cup sugar (more or less to your taste)
- 3 tablespoons quick-cooking tapioca
- 1 to 2 tablespoons lemon juice

1. Tear off a 24-inch piece of heavy-duty aluminum foil or several 24-inch sheets of plastic wrap and use to line an 8-inch pie pan. Let 6 inches of foil or wrap hang over each side of pan.
2. Stir the fruit and ascorbic acid together, then stir in the remaining ingredients.
3. Turn the fruit mixture into a lined pie pan. Fold the foil or wrap lightly over filling.
4. Freeze solid.
5. Wrap it tightly with foil or wrap that is loose over the top; seal with freezer tape, label and store in the freezer.
6. To Bake: Prepare pastry for double crust pie. Roll out half and fit it into a 9-inch pie plate or pan. Unwrap the frozen filling and put it into pastry. Dot with 1½ tablespoons butter or margarine. Roll out the remaining pastry for the top crust; put it in place, seal and crimp edges. Cut vents in the top. Bake in preheated 425°F oven 1 hour.

Freezing Vegetables

Preparing vegetables for the freezer is just like preparing them for dinner: pick for perfection, wash well, trim and cut or slice. For freezing, precook (blanch) the vegetables very briefly, then quickly cool them to pack in containers, bags or boilable pouches; seal, label and freeze.

Blanching

No matter what your neighbor may say, blanching vegetables before freezing is ABSOLUTELY NECESSARY. Blanched vegetables keep their color, flavor and freshness in freezer storage; unblanched vegetables probably will not. Blanching is a simple time saving step and skipping it is foolish.

Basic Equipment

Since you are going to blanch, you will need a blancher. You can find one in the housewares section of most stores. It is a large pot, with a cover and a perforated insert or basket insert for lifting vegetables out of the boiling water. A blancher will not gather dust when you are not freezing vegetables; use it as a spaghetti cooker or soup pot, a steamer, even a deep fat fryer. You can make your own blancher from a large (6- to 8-quart) pot with a cover and something (a colander, sieve, deep frying basket, or cheesecloth bag) to hold the vegetables in the boiling water. Besides a blancher, have this equipment clean and ready:

Sharp paring knife
Chopping knife
Colander, sieve, strainer or paper towels
Stiff vegetable scrubbing brush
Teakettle for extra boiling water
Ricer, food mill or blender for mashing or pureeing
Containers with tight-fitting lids: plastic freezer containers, shortening or coffee cans, freezer cans or jars with wide mouths
Or, bag containers: polyethylene or heavy-duty plastic bags, or boilable pouches
String, rubber bands, pipe cleaners or twist ties to fasten bags
Shallow tray, cookie sheet, jelly roll pan for tray packing
Freezer tape
Labels

Basic Ingredients

Choose vegetables that are tender, ripe but just barely ready to eat and as fresh as possible. Slightly under-mature vegetables are better for freezing than those that are past their prime. For peak flavor, rush vegetables from the garden to the freezer within two hours. If you cannot make it in that time, keep the vegetables refrigerated until ready to prepare.

Ice

Since quick-cooling is an important part of preparing vegetables for freezing, you will need plenty of ice at hand to keep the cooling water really cold. Estimate one pound of ice for each pound of vegetables you are going to freeze. Empty the ice cube trays into a large, heavy-duty plastic or brown grocery bag whenever you think of it and store them in the freezer

Steps 2 and 3: Select tender, young, fresh vegetables and wash them carefully.

to build up an ice reserve. Or, whenever you empty a cardboard milk carton, rinse it out, fill it with water, and freeze. (If you make your own ice cream, you will find these ice-reserve ideas useful, too.)

The kitchen sink is a favorite spot for holding ice water to cool vegetables, but if you want it free for other uses, put the ice water in a plastic dishpan or other large, clean container.

Basic Steps

1. Check your freezer's size and estimate how much food you can freeze in a 24-hour period (3 pounds for each cubic foot of space), then check the recipe for the vegetable you want to freeze to see how much fresh will yield

Step 4: Fill blancher with 1 gallon water (2 gallons for greens); place it on high heat.

how much frozen. Figure how much you should pick or buy for a single preparation session.

2. Select tender, young, fresh vegetables that are unblemished.
3. Wash vegetables well in plenty of cold water, scrubbing with a brush whenever necessary. Lift vegetables out of water to drain. Do not let them stand in water any longer than necessary because they lose nutrients. Sort by size and handle like sizes together for even heating and cooling.
4. Put 1 gallon water (2 gallons for greens) in the blancher, cover and place it on high heat.
5. Prepare the sink or dishpan full of ice cubes and ice water.
6. Cut or prepare the vegetable, about 1 pound or 4 cups at a time, as the recipe directs.

Step 7: Put 1 pound prepared vegetables in blancher's insert, lower into boiling water. Start timing.

Container Pack. Put vegetables gently but firmly into containers, leaving head space as directed in recipe.

7. Put 1 pound prepared vegetables in the blancher's insert and lower it into rapidly boiling water. Keep the heat high. Cover and begin timing immediately.
8. When the time is up, remove the cover, lift the blancher's insert up out of the blancher for a few seconds to drain, then immediately put the insert of vegetables into ice water.
9. Keep vegetables in the ice water for about the same length of time as in the boiling water, or until cold.
10. Lift the vegetables from the ice water and drain them well in a colander, sieve or on paper towels.
11. Pack into containers, bags, or freeze on trays.
12. Label each package with food and date.
13. Freeze, following the Basic Rules for Freezing.
14. Vegetables will keep for 12 to 18 months at 0°F.

Bag Pack. Put vegetables gently but firmly into bags, leaving several inches of bag to twist for seal.

Packing

Containers: Pack vegetables gently but firmly into containers to press the air out. Leave ½ inch of head space (unless the recipe directs otherwise). Wipe the sealing area clean with a damp cloth. Put on the lid and seal.
Bags: Pack vegetables gently but firmly into bags, leaving several inches of bag to twist for the seal. Press out the air and twist the top of the bag, leaving ½ inch of head space. Double over the twisted area and fasten with string, rubber band, pipe cleaner or twist tie.
Pouches: Follow instructions that come with the heat sealer for sealing boiling pouches or bags.
Tray: Spread vegetables in single layer on a cookie sheet, jelly roll pan or shallow tray. Freeze just until solid. Check

After blanching vegetables cool in ice. These asparagus are left whole for a tray pack.

Tray Pack. Spread vegetables in single layer on a cookie sheet, jelly roll pan or shallow tray.

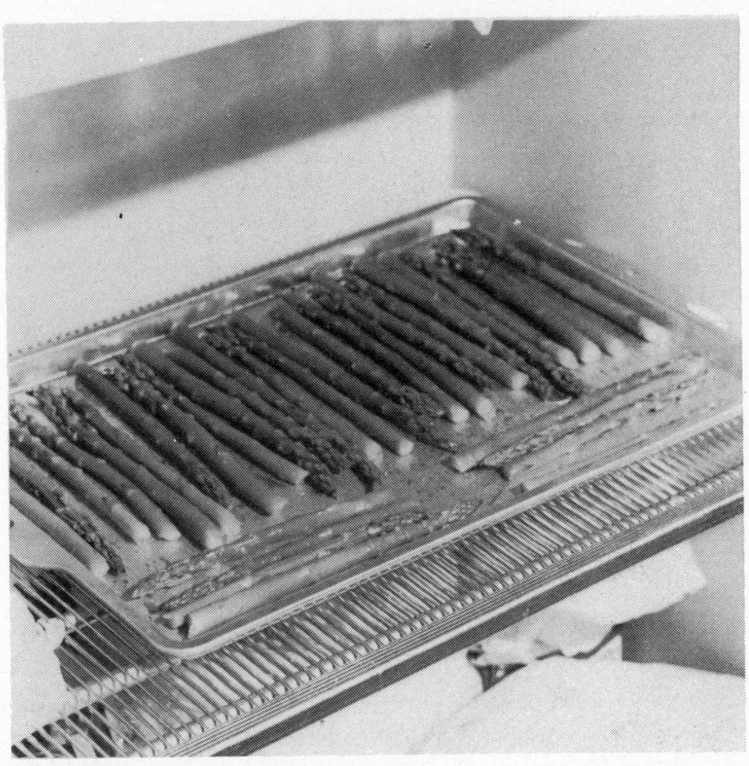

Tray Pack. Freeze just until solid. Note time for future reference. Transfer to containers for freezer storage.

after an hour in the freezer, then note how long it takes to freeze the vegetable so you will have the time for future reference. As soon as the vegetables are solid, transfer them to containers, bags or pouches, seal and store in the freezer.

Blanching Water

Blanching water must be boiling when you lower the vegetables into it. You can use the same water for several batches of vegetables, adding additional boiling water from a teakettle to replace water lost through evaporation. If you wish, change the water when it becomes cloudy. Keep a second pot or large teakettle boiling so you will not be delayed when the time comes to change the blanching water.

Be sure to add ice to ice water frequently, so it stays as cold as possible.

To Cook Frozen Vegetables

Do not thaw frozen vegetables. For each pint of frozen vegetables, heat ¼ to ½ cup water to boiling in a small or medium saucepan. Add the vegetables and keep them over high heat until the water returns to a boil. Break vegetables into individual pieces with a fork as they heat. When the water boils again, cover the pan, reduce the heat and simmer just until the vegetable is tender. Recipes give you approximate times. Begin timing when the cooking water returns to a boil. When crisp-tender, add butter or margarine, salt, pepper or other seasonings to taste.

For tray-frozen vegetables, just pour out the amount you need, using a little more than ½ cup water if you pour out more than 2 cups of vegetables. Cook as above.

For pouched vegetables, follow the cooking directions that come in the heat-sealer's instruction book.

Butter Sauce For Pouched Vegetables

For each 1 to 1½ cups vegetables in small boilable pouches add 2 tablespoons water, 1 tablespoon butter, ¼ to ½ teaspoon salt and dash oregano, basil, savory, chervil, tarragon, thyme, sage or marjoram. Seal, label and freeze. Heat as the manufacturer directs.

Recipes

ASPARAGUS

1 to 1½ pounds = 1 pint, frozen
1 crate (24 pounds) = 15 to 22 pints, frozen

1. Choose very fresh stalks that are brittle and beautifully green. The tips should be compact and tight.

2. Wash well and sort by size.
3. Break off the tough ends. Leave the stalks whole or cut into 1- or 2-inch lengths.
4. Blanch small stalks for 1½ to 2 minutes, medium stalks 2 to 3 minutes, large stalks 3 to 4 minutes. Cool; drain well.
5. Pack into containers, alternating tip and stem ends.
6. Seal, label and freeze.
7. Cook frozen asparagus 5 to 10 minutes.

BEANS, GREEN AND WAX

²/₃ to 1 pound fresh = 1 pint frozen
1 bushel (30 pounds) fresh = 1 pint frozen

1. Choose young, tender beans that snap easily. The beans in the pods should not be fully formed.
2. Wash well, snip the ends and sort by size.
3. Leave the beans whole, cut in even lengths or cut French-style (lengthwise).
4. Blanch 2 to 3 minutes, depending on the size of the beans. Cool; drain well.
5. Tray freeze or pack into containers, leaving ½ inch of head space. Seal, label and freeze.
6. Cook frozen cut beans 12 to 18 minutes, French-style beans 5 to 10 minutes.

BEANS, ITALIAN

Fresh to frozen yields will be about the same as for green beans.

1. Choose wide, flat beans that are tender, meaty and stringless.
2. Wash well, snip off the ends and cut or break them into 1½-inch pieces.
3. Blanch 3 minutes. Cool; drain well.
4. Tray freeze or pack into containers, leaving ½ inch of head space. Seal, label and freeze.
5. Cook frozen beans about 10 minutes.

BEANS, LIMA

2 to 2½ pounds, in pods = 1 pint, frozen
1 bushel (32 pounds), in pods = 12 to 16 pints, frozen

1. Choose well-filled pods of young, green, tender beans.
2. Shell and sort by size.
3. Blanch small beans 2 minutes, medium beans 3 minutes, large beans 4 minutes. Cool; drain well.
4. Tray freeze or pack into containers, leaving ½ inch of head space.
5. Cook large frozen beans 6 to 10 minutes, baby limas 15 to 20 minutes.

SOYBEANS

Fresh to frozen yields will be about the same as for Lima Beans.

1. Choose well-developed, well-filled pods with green beans.
2. Wash the pods well.
3. Blanch beans in pods 5 minutes. Cool; drain well.
4. Squeeze beans out of the pods and pick over, discarding any bad beans.
5. Tray freeze or pack into containers, leaving ½ inch of head space. Seal, label and freeze.
6. Cook frozen soybeans 10 to 20 minutes.

BEETS

1¼ to 1½ pounds, without tops = 1 pint, frozen
1 bushel (52 pounds), without tops = 35 to 42 pints, frozen

1. Choose young, tender beets no more than 3 inches in diameter.

2. Remove tops, leaving ½ inch of stem. Wash.
3. Cook in boiling water to cover, 25 to 30 minutes for small beets, 45 to 50 minutes for medium.
4. Cool and peel, then slice, cube or dice.
5. Pack into containers, leaving ½ inch of head space. Seal, label and freeze.
6. Cook frozen beets just until heated through.

BEET GREENS

1 to 1½ pounds greens = 1 pint, frozen
15 pounds greens = 10 to 15 pints, frozen

1. Choose young, tender leaves.
2. Wash well, in several changes of water.
3. Remove tough stems and bruised leaves.
4. Blanch each pound of greens in 2 gallons boiling water for 2 minutes. Cool; drain well.
5. Pack into containers, leaving ½ inch of head space. Seal, label and freeze.
6. Cook frozen beet greens 6 to 12 minutes.

BROCCOLI

1 pound = 1 pint frozen
1 crate (25 pounds) = 24 pints frozen

1. Choose compact, dark green heads. Stalks should be tender, not woody.
2. Wash well, then soak heads of broccoli in ½ cup salt per gallon of water for ½ hour, to get rid of bugs. Rinse in fresh water; drain.
3. Cut stalks to fit into the containers. Split stalks so that the heads are about 1 to 1½ inches in diameter.
4. Blanch 4 minutes. Cool; drain well.
5. Pack into containers. No head space is necessary. Seal, label and freeze.
6. Cook frozen broccoli 5 to 8 minutes.

BRUSSELS SPROUTS

1 pound = 1 pint frozen
4 quart boxes = 6 pints frozen

1. Choose firm, compact, bright green heads.
2. Wash thoroughly, then soak in salt water as for broccoli. Rinse in fresh water; drain. Sort by size.
3. Blanch small sprouts 3 minutes, medium sprouts 4 minutes, large sprouts 5 minutes. Cool; drain well.
4. Pack into containers. No head space is necessary. Seal, label and freeze.
5. Cook frozen sprouts 4 to 9 minutes.

CABBAGE OR CHINESE CABBAGE

Frozen cabbage can only be used as a cooked vegetable.

1 to 1½ pounds = 1 pint, frozen

1. Choose fresh, solid heads.
2. Trim the coarse outer leaves.
3. Cut medium coarse shreds, thin wedges, or separate into leaves.
4. Blanch 1 to 1½ minutes. Cool; drain well.
5. Pack into containers, leaving ½ inch of head space. Seal label and freeze.
6. Cook frozen cabbage about 5 minutes, or thaw leaves to use for stuffed cabbage or cabbage rolls.

CARROTS

1¼ to 1½ pounds without tops = 1 pint, frozen
1 bushel (50 pounds) without tops = 32 to 40 pints, frozen

1. Choose tender, small to medium, mild-flavored carrots.

2. Remove tops, wash well, peel.
3. Cut in cubes, slices or lengthwise strips, or leave small carrots whole.
4. Blanch small whole carrots 5 minutes; blanch diced or sliced carrots and lengthwise strips 2 minutes. Cool; drain well.
5. Pack into containers, leaving ½ inch of head space. Seal, label and freeze.
6. Cook frozen carrots 5 to 10 minutes.

CAULIFLOWER

1⅓ pounds = 1 pint, frozen
2 medium heads = 3 pints, frozen

1. Choose compact, white, tender heads.
2. Break into flowerets about 1 inch in diameter.
3. Wash well, then soak in salt water as for broccoli. Rinse with fresh water; drain.
4. Blanch 3 to 4 minutes. Cool; drain well.
5. Pack into containers. No head space is necessary. Seal, label and freeze.
6. Cook frozen cauliflower 5 to 8 minutes.

CELERY

Use frozen celery in cooked dishes only.

1 pound fresh = 1 pint, frozen

1. Choose crisp tender stalks without coarse strips.
2. Wash well, trim and cut into 1-inch lengths.
3. Blanch 3 minutes. Cool; drain well.
4. Tray freeze or pack into containers, leaving ½-inch head space. Seal, label and freeze.
5. Cook frozen celery with other vegetables in soups, stews and casseroles.

CORN, WHOLE KERNEL

2 to 2½ pounds, in husks = 1 pint, frozen
1 bushel (35 pounds) in husks = 14 to 17 pints, frozen

1. Choose well-developed ears with plump tender kernels and thin, sweet milk. Press a kernel with a thumbnail to check the milk. Corn must be fresh to freeze.
2. Husk, remove the silk and trim the ends. Sort by size.
3. Blanch small ears 5 minutes, medium 8 minutes, large ears 10 minutes. Cool; drain well.
4. Cut the kernels from the cob at ⅔ the depth of the kernel.
5. Tray freeze or pack in containers, leaving ½ inch of head space. Seal, label and freeze.
6. Cook frozen corn 3 to 4 minutes.

CREAM-STYLE CORN

1. Prepare as for Whole Kernel Corn, but cut the kernels from cob at ½ the depth of the kernel. Then scrape the cob with the back of a knife to remove the milk and heart of the kernel.
2. Pack as for Whole Kernel Corn. Seal, label and freeze.
3. Cook as for Whole Kernel Corn.

CORN ON THE COB

1. Choose corn and prepare as for Whole Kernel Corn.
2. Blanch small ears (1¼ inches or less in diameter) 7 minutes; medium ears (1¼ to 1½ inches diameter) 9 minutes; large ears (more than 1½ inches) 11 minutes. Cool thoroughly; drain well.
3. Pack in containers or wrap the ears individually or in family-size amounts in freezer paper, plastic wrap or foil. Seal, label and freeze.

4. Thaw frozen corn on the cob before cooking, then cook about 4 minutes.

GREENS (Chard, collards, kale, mustard, turnip and spinach)

1. Follow directions for Beet Greens, except blanch the collards for 3 minutes; for very tender spinach blanch only 1½ minutes.
2. Cook frozen greens 8 to 15 minutes.

KOHLRABI

1¼ to 1½ pounds = 1 pint, frozen

1. Choose young, tender, small to medium kohlrabi.
2. Cut off the tops and roots and wash well.
3. Peel, leave whole or slice ¼ inch thick or dice into ½-inch cubes.
4. Blanch whole kohlrabi 3 minutes; blanch diced or sliced kohlrabi 1 to 2 minutes.
5. Tray freeze or pack in containers, leaving ½ inch of head space. Seal, label and freeze.
6. Cook frozen kohlrabi 8 to 10 minutes.

MUSHROOMS

1 to 2 pounds = 1 pint, frozen

1. Choose young, firm, medium mushrooms with tightly closed caps.
2. Wash well, trim off ends and sort by size. Slice larger mushrooms, if desired.
3. Cook mushrooms in a frypan with a small amount of butter until almost done. Cool.
4. Pack in containers, leaving ½ inch of head space. Seal, label and freeze.
5. Cook frozen mushrooms just until heated through.

OKRA

1 to 1½ pounds = 1 pint, frozen

1. Choose young, tender green pods.
2. Wash well. Cut off the stems but do not cut open the seed cells.
3. Blanch small pods 3 minutes, large pods 4 minutes. Cool; drain well.
4. Slice crosswise in ½- to ¾-inch slices or leave whole.
5. Pack into containers, leaving ½-inch head space. Seal, label and freeze.
6. Cook frozen okra about 5 minutes.

ONIONS

1 to 3 medium onions = 1 pint, frozen chopped onions

1. Choose the best quality, fully mature onions. Freeze only chopped or sliced onions, store whole onions in a cool, dry place.

2. Peel, wash and chop or slice.
3. Package onions in recipe-size amounts or tray freeze them and then pack or bag the onions in recipe-size amounts so you can pour out what you need. Onions do not need to be blanched, but they should not be stored in the freezer longer than 3 to 6 months at 0°F.

PARSNIPS

Fresh to frozen yields will be about the same as for carrots.

1. Choose small to medium, tender, not woody parsnips.
2. Remove tops, wash, peel and cut in ½-inch cubes or slices.
3. Blanch 2 minutes. Cool; drain well.
4. Pack in containers, leaving ½ inch of head space. Seal, label and freeze.
5. Cook frozen parsnips about 10 to 12 minutes.

PEAS

2 to 2½ pounds = 1 pint, frozen
1 bushel (30 pounds) = 12 to 15 pints, frozen

1. Choose fresh, bright green, plump, firm pods with sweet and tender peas.
2. Wash pods and shell the peas.
3. Blanch 1½ to 2 minutes. Cool; drain well.
4. Tray freeze or pack peas in containers, leaving ½ inch of head space. Seal, label and freeze.
5. Cook frozen peas 5 to 10 minutes.

PEPPERS, GREEN OR HOT

$^2/_3$ pound fresh = 1 pint frozen

1. Choose shapely, firm, well-colored peppers.
2. Cut out the stems, remove seeds.
3. Leave whole, halve, slice or dice. There is no need to blanch peppers.
4. Tray freeze or pack into containers. No head space is necessary. Seal, label and freeze.
5. Add frozen chopped pepper to uncooked or cooked dishes, or cook about 5 minutes.

PIMIENTOS

1. Choose firm, crisp pimientos with thick walls.
2. Roast pimientos in 400°F oven 3 to 4 minutes.
3. Rinse in cold water to remove charred skins. Drain.
4. Pack into containers, leaving ½ inch of head space. Seal, label and freeze.
5. Chop frozen pimientos and add them to cooked dishes.

POTATOES, BAKED

1. Bake the potatoes, then let them cool. Scoop out the potato and mash. Season and stuff the mashed potato back into the shells, if desired.
2. Tray freeze, then wrap the shells individually. Seal, label and freeze.
3. Cook by unwrapping and reheating in a 400°F oven until hot, about 30 minutes.
4. Freeze leftover baked potatoes unstuffed and then slice or cube them for creamed or scalloped potatoes, potato salad, or American fries.

POTATOES, FRENCH FRIED

1. Choose large, mature potatoes.
2. Wash, peel and cut them in sticks about ½-inch thick.
3. Rinse well in cold water; drain. Pat dry.
4. Fry small amounts at a time in deep fat (360°F) for 5 minutes or until tender but not brown.
5. Drain well on paper towels. Cool.
6. Pack into containers. No head space is necessary. Seal, label and freeze.
7. Cook frozen French fries in deep fat (375°F) until browned, or spread in single layer on a cookie sheet and heat in a 450°F oven 5 to 10 minutes, or until browned.

PUMPKIN

3 pounds = 2 pints, frozen

1. Choose finely-textured, ripe and beautifully colored pumpkins.
2. Wash, cut in quarters or small pieces and remove seeds.
3. Cook in boiling water, steam, pressure-cook or oven-cook until tender.
4. Scoop the pulp from the skin and mash in a saucepan or press through a ricer, sieve or food mill into a saucepan.
5. Cool by putting the saucepan in ice water and stirring the pumpkin occasionally until cold.
6. Pack into containers, leaving 1 inch of head space. Seal, label and freeze.
7. Cook frozen pumpkin just until heated through.

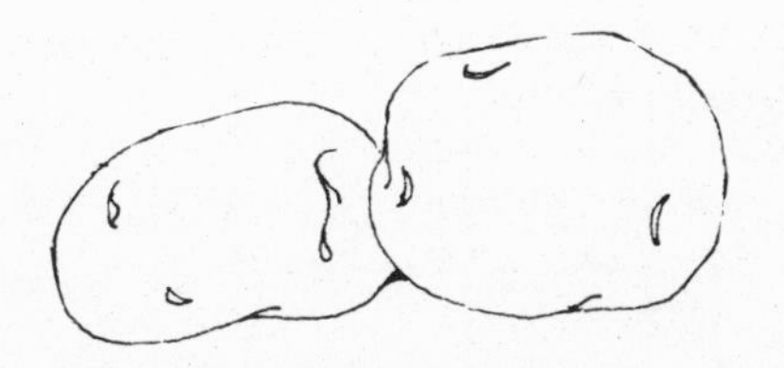

PUMPKIN PIE MIX

1. Prepare the pumpkin as above.
2. Combine measured amounts of mashed pumpkin with the remaining ingredients in your favorite recipe (omit cloves if used and add after thawing because their flavor changes).
3. Pour the pumpkin into containers in single pie amounts. Seal, label and freeze.
4. Thaw, add cloves, turn into a pastry shell and bake as your recipe directs.

RUTABAGA

1¼ to 1½ pounds = 1 pint frozen

1. Choose young, tender, medium rutabaga.
2. Cut off tops, wash and peel. Cut in ½-inch cubes.
3. Blanch for 3 minutes. Cool; drain well.
4. Tray freeze or pack into containers, leaving ½ inch of head space. Seal, label and freeze.
5. Cook frozen rutabaga 12 to 15 minutes.
6. For Mashed Rutabaga: proceed as above, cooking cubed rutabaga in just enough water to cover until tender, instead of blanching. Drain, then mash in saucepan. Cook as for Pumpkin. Pack in containers, leaving 1 inch of head space. Seal, label and freeze. Cook mashed rutabaga just until heated through.

SQUASH, SUMMER

1 to 1¼ pounds = 1 pint, frozen
1 bushel (40 pounds) = 32 to 40 pints, frozen

1. Choose young, tender squash.
2. Wash, cut off the ends and slice the squash ½ inch thick.
3. Blanch 3 minutes. Cool; drain well.
4. Tray freeze or pack into containers, leaving ½ inch of head space. Seal, label and freeze.
5. Cook frozen summer squash about 10 minutes.

SQUASH, WINTER

3 pounds fresh = 2 pints frozen

1. Choose firm, mature squash with hard skins.
2. Wash, halve or quarter the squash. Bake, simmer or pressure-cook it until tender.
3. Scoop pulp from the skin and mash the pulp in a saucepan or press it through a sieve, ricer or food mill into a medium saucepan.
4. Cool by placing the saucepan in ice water and stirring the squash occasionally until cold.
5. Pack into containers, leaving 1 inch of head space. Seal, label and freeze.
6. Cook frozen squash just until heated through.

SWEET POTATOES

3 to 4 medium potatoes = 1 pint, mashed frozen

1. Choose medium or large mature potatoes that have been cured.
2. Wash well and sort by size.
3. Simmer, bake or pressure-cook the potatoes until tender. Cool.
4. Peel, then halve, slice or mash.
5. Dip the halves or slices in 1 quart water plus ½ cup lemon juice to prevent discoloration. Mix mashed sweet potatoes with 2 tablespoons orange or lemon juice for each quart mashed potatoes.
6. Pack into containers, leaving ½ inch of head space for halves or slices, 1 inch for mashed. Seal, label and freeze.
7. Cook frozen sweet potatoes just until heated through.

TOMATOES

1 bushel (53 pounds) = 30 to 40 pints, frozen stewed

1. Choose ripe, firm, red tomatoes, free of blemishes.
2. Wash well.
3. Whole Tomatoes: Remove the stems after washing. Wrap each tomato in plastic wrap or a small plastic bag, freeze. To use in cooked dishes, run under lukewarm water for a few seconds to loosen the skin, then remove the skin. Add tomato along with other ingredients and cook.
   Stewed Tomatoes: Dip whole, washed tomatoes in boiling water 2 minutes to loosen skins. Peel and core. Cut in quarters or pieces; simmer 10 to 20 minutes or until tender. Cool. Pack in containers, leaving ½ inch of head space. Seal, label and freeze.
   Puree: Peel tomatoes as for Stewed Tomatoes. Core and cut them in quarters into a blender container. For each 4 medium tomatoes, add ½ onion, chunked; 1 green pepper (seeded, stemmed and quartered); 1 tablespoon salt and 1 tablespoon sugar. Blend on low or medium speed until the onion and pepper are chopped. Pack the puree in containers leaving ½ inch of head space. Seal, label and freeze.

TURNIPS

1¼ to 1½ pounds = 1 pint, frozen

1. Choose young, tender, small to medium-size turnips.
2. Prepare, blanch and freeze as for Parsnips.

VEGETABLE PUREES

Freeze a puree to have on hand for casseroles, cream soups, special diets or for baby food.

1. Choose and prepare vegetables as directed in specific recipes.
2. Cook in just enough boiling water to cover, or steam, bake or pressure cook until the vegetables are tender.
3. Mash them in a saucepan, or press through a sieve, ricer, or food mill into a saucepan, or puree in blender and pour into a saucepan.
4. Cool by putting the saucepan in ice water and stirring the puree occasionally until cold.
5. Pack in recipe-size amounts in containers, leaving 1 inch of head space. Seal, label and freeze.
6. Thaw in the refrigerator or in a double boiler. For Baby Food: Pour puree into ice cube trays and freeze solid. Transfer to plastic bags, seal and store in the freezer. Thaw in a custard cup over boiling water or in a microwave oven.

HERBS

Basil, Chervil, Chives, Dill, Fennel, Marjoram, Mint, Parsley, Rosemary, Sage, Tarragon, Thyme.

1. Pick fresh, perfect herb sprays, wash well and pat dry.
2. Pack recipe-size amounts in small plastic bags or packets made from plastic wrap, freezer paper or foil. Seal.
3. Staple these individual packets to a piece of cardboard, label the cardboard and freeze. Or, pack several packets in a large envelope or rigid freezer container or large plastic bag, seal, label and freeze.
4. Snip frozen herbs directly into foods to be cooked, using them just as you would fresh herbs. Frozen herbs tend to become dark and limp; they do not make very beautiful garnishes, but their flavor is just as good as fresh.
   Bouquet Garni: Tie together sprigs of several different herbs, pack as above. Add the whole bouquet to foods to be cooked.

Freezing Meat, Poultry And Fish

Meat is probably the most expensive item you store in your freezer, so wrapping it and storing it properly is a matter of dollars and cents.

If you are thinking of buying meat in quantity, shop wisely and carefully. There are reputable and disreputable dealers and plans; sometimes it is hard to tell one from another until it is too late. If in doubt about a freezer "plan" or "program," or beef half or quarter dealer, check with the local Better Business Bureau. The Federal Trade Commission has an informative booklet, "Bargain?" Freezer Meats, Consumer Bulletin

#5. You can probably get a copy at your county extension office. It is worth reading.

Basic Equipment

The equipment for freezing meat is the same as Basic Equipment for Freezing, with emphasis on the following:

Sharp knife
Moisture-vapor-proof wrapping: freezer paper, heavy-duty aluminum foil, heavy-duty plastic wrap, heavy-duty plastic bags
Ties for bags: rubber bands, twist ties, pipe cleaners
Freezer tape
Grease pencil or felt tip marker
Labels
Shallow tray, jelly roll pan or cookie sheet for tray freezing

Basic Ingredients

What meat to buy? That depends on you and your family. If you are a hamburger and pot roast gang, it does not make much sense to buy a quarter or half steer and wind up with many cuts you really do not care for. It is more sensible to buy your favorite cuts when they are on sale or special. Watch supermarket ads carefully, or cultivate the friendship of a market meat man who will let you know when specials are due. When big roasts are on special you can usually divide them into steaks, or chops, and still have a small roast left in the middle. Ask the meat man for guidance in order to have some meat to eat right away plus more to freeze for other meals.

Whatever meat you choose, be sure it is fresh and of good quality. Look for the "U.S. Inspected and Passed" stamp as a guarantee of wholesomeness. Packers who do not ship interstate do not have to be federally inspected, but usually must pass state or city inspections, so look for a state or city stamp if the U.S. one is missing.

Packaging

Trim the excess fat from the meat before wrapping it for the freezer. Remove the bones, if you wish, or at least cover sharp ends with freezer paper or foil so they will not pierce wrappings.

Package meats in the amount you will use at one meal. Meat must be wrapped in moisture-vapor-proof material (freezer paper, heavy-duty aluminum foil, heavy plastic wrap, heavy plastic bags) and sealed with freezer tape. Store wrappings are good for about a week in the freezer. For longer storage, unwrap the meat and rewrap or overwrap the store wrappings. Large, heavy plastic bags, tightly closed or heat sealed, are good meat holders, too. Wrap meat using a Drugstore Wrap or Butcher Wrap. Freeze as directed in Freezing Rules.

Tray Freezing

Individual pieces of meat, such as patties, steaks and chops, can be tray frozen. Put two sheets of freezer paper or plastic wrap between each piece of meat, then pack the frozen meat in plastic bags or wrap. Seal and store in the freezer.

Storage

Meat	Freezer Storage at 0°F
Beef	*6 to 12 months*
Veal	*6 to 9 months*
Lamb	*6 to 9 months*
Pork	*3 to 6 months*
Ground beef, lamb and veal	*3 to 4 months*
Ground pork	*1 to 3 months*
Variety meats (liver, heart, etc.)	*3 to 4 months*
Whole or sliced ham	*2 months*
Fresh pork sausage	*2 months*
Bacon	*1 month*
Hot dogs	*1 month*

Thawing

Most frozen meat does not need to be thawed before cooking,

but you do need to add a third to half again as much cooking time if you start with frozen meat. Always use low, even temperatures for roasting (325°F). When the meat has thawed enough to be soft, put a meat thermometer in the center of the largest muscle, away from the bone and fat, and use it to determine when the meat is done.

The only meats that must be thawed before cooking are those that you are going to shape or stuff, those that are to be coated with breading or batter, or those that are to be deep fat fried.

Always thaw meats in the refrigerator, never at room temperature. If you have a microwave oven, follow the manufacturer's directions for thawing (defrosting).

Refrigerator thawing does take time, so you have to plan ahead, usually transferring the meat from the freezer to the refrigerator the night before.

Poultry

There are very few freezers in this country that do not hold several fryers, bought when prices were low. And, many a "freezer" cook has discovered what a bargain turkey can be.

Equipment and Ingredients

You need the same equipment as for Freezing Meat.

Look for fresh, inspected, chilled or frozen poultry, then hurry it home from the market. Get frozen birds into the freezer fast. Unwrap fresh birds, rinse them with cold water, pat dry, then package for the freezer.

Packaging

Pieces: Freeze cut up birds on a tray, then seal the frozen pieces in bags and freezer store. Or you can wrap individual pieces in plastic wrap, freezer paper or plastic bags. Wrap or bag, then seal, label and freeze.

Whole chickens are often cheaper than pieces, so save some pennies by buying whole birds, then cut them up yourself.

Always work on a clean cutting board, with a clean knife

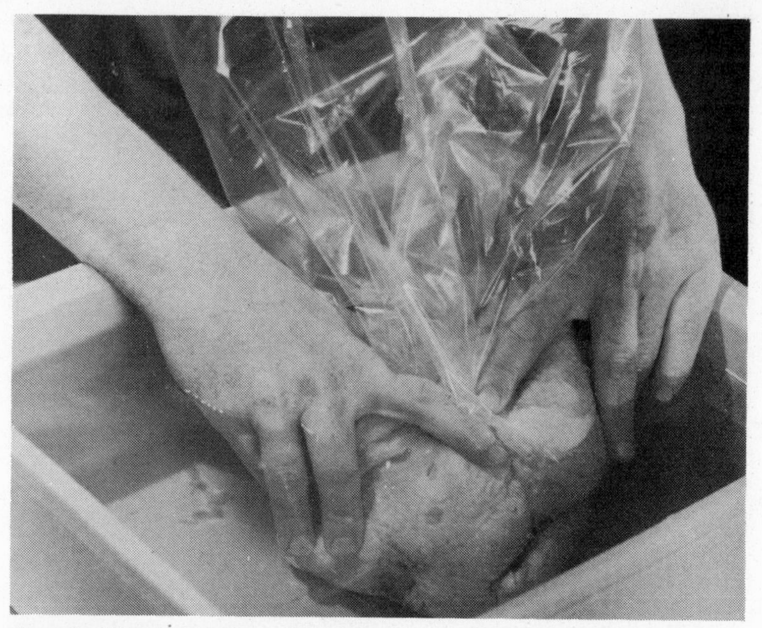

Freezing Whole Chicken. Put the bird in a plastic bag, lower into cold water to get air out.

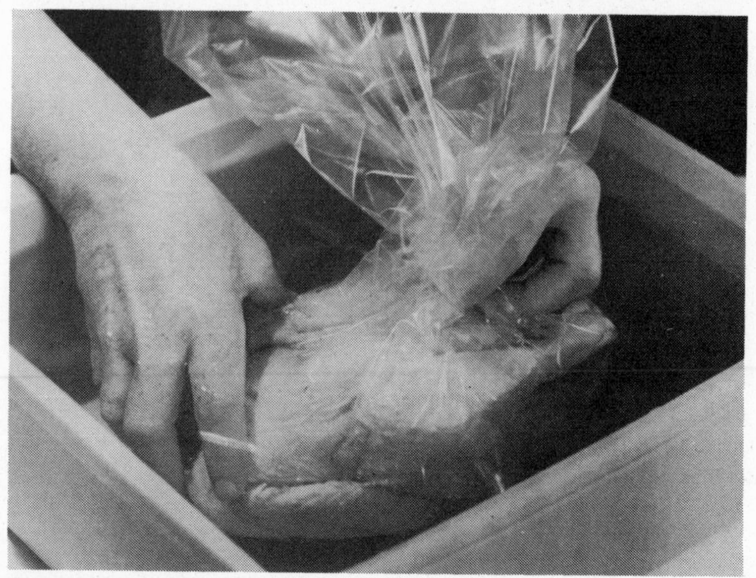

Do not let any water into the bag. Squeeze out air bubbles and twist the top.

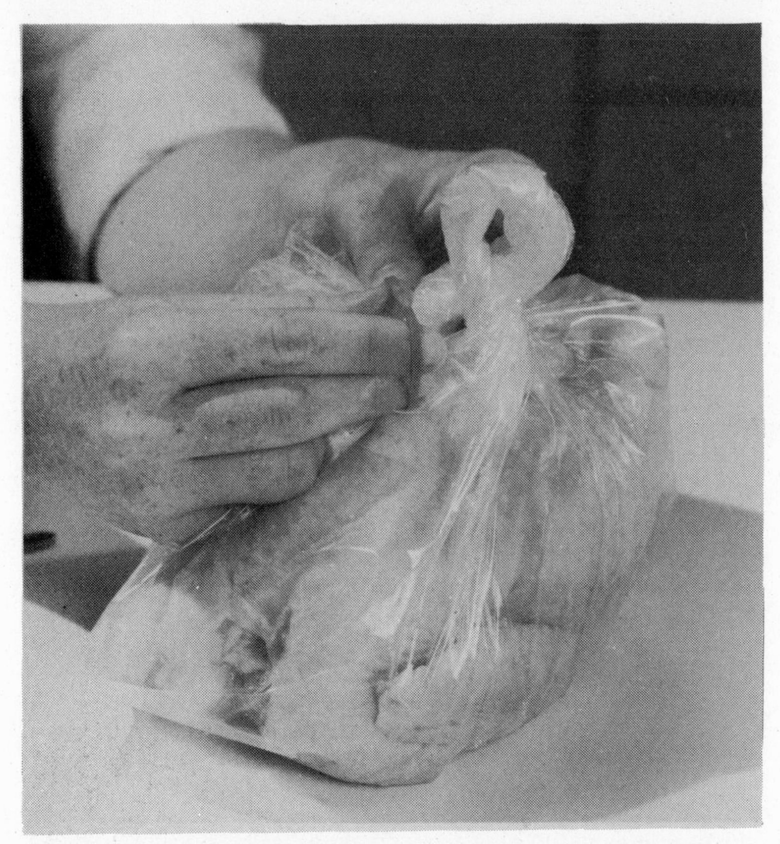

Double over the twisted top and seal with string, rubber band, pipe cleaner or twist tie. Label and freeze.

and clean hands, and work quickly so poultry stays cold. Scrub all equipment well when you are through cutting, using hot suds followed by a good hot rinse.

Whole: Remove the giblets and neck and rinse. Tray freeze the giblets separately; seal in bags when frozen solid. Rinse the birds inside and out; drain well. Wrap in freezer paper, foil or plastic wrap, using the Drugstore or Butcher Wrap. Or put the whole bird in a plastic bag. To get as much air out as possible, lower the bird in a plastic bag into a sink of cold water. Do not let any water in the bag. Twist the bag's top, double over the twisted area and seal with string, rubber band, pipe cleaner or twist tie. Label and freeze.

Storage

Poultry	**Freezer Storage at 0°F**
Chicken	*12 months*
Turkey	*12 months*
Duck	*6 months*
Geese	*6 months*
Giblets	*3 months*

Thawing

Aways thaw poultry, unopened, in the refrigerator (or in a microwave oven, following the manufacturer's directions). Never thaw at room temperature. When thawed, cook it as soon as possible. Poultry pieces can be baked or simmered without thawing. Pieces to be deep fried must be thawed, as must whole birds.

Cooking For The Freezer

When chickens or turkeys are priced right, freeze some raw, but cook some to start making your own convenience foods. Simmer chickens in enough water to cover, adding an onion, some celery, seasonings of your choice. Chill the chicken in the broth. When cold, remove the fat, lift the chicken from the broth. Package fat and broth separately to freeze for soups and sauces. Cut the chicken into slices, cubes or chunks and freeze in recipe-size amounts for casseroles, crepe fillings, chicken a la king, salads and sandwiches. Roast capons, roasting chickens or turkeys. Cut the meat from the carcass and package as for chicken. Simmer the carcass in water to cover, strain the stock and freeze for soup, sauces and gravy. You can also fry chicken in large quantities, serve some for dinner, and cool, wrap, seal, label and freeze the rest.

Fish

If you plan to freeze a catch from a stream instead of a market, be sure the fisherman (or woman) has ice and a cooler, ice chest or some other portable chiller to chill the fish and keep it cold until he (she) gets home.

Equipment and Ingredients

You will need the equipment listed under Basic Equipment for Freezing.

Clean and prepare fish as if you would for the table, gutting, removing head, fins and tail if desired. A quick dip in salt water ($^{2}/_{3}$ cup salt dissolved in 1 gallon cold water) for 30 seconds helps preserve fish. Drain well.

Packaging

Whole: Seal fish in plastic bags or wrap in foil, plastic wrap or freezer wrap, seal, then bag it in a large plastic bag. Label and freeze. You can also put a fish in a rigid freezer container, empty milk carton, heavy plastic bag or loaf pan, cover with water and freeze solid. This ice-block method keeps the fish really fresh, but does take up more freezer space than ordinary sealing. Another preparation method for whole fish is ice glazing. Chill trays or shallow pans in the freezer so they will be cold. Dip the fish in very icy water, then tray freeze on chilled trays until solid. Quickly transfer the ice-glazed fish to plastic bags or containers. Seal and store in the freezer.

Fish Steaks or Fillets: After dipping in salt water, drain, then wrap each piece separately in foil, plastic wrap or freezer paper. Package the pieces in family-size amounts in bags or wrap them. Seal, label and freeze.

Storage and Thawing

Fish can be stored in a freezer at 0°F 6 to 9 months. Thaw fish in the refrigerator and cook as soon as possible once it is thawed.

Shellfish

Choose very fresh, live shellfish; prepare them quickly to preserve freshness.
Lobsters and Crabs: Wash them well; drop them live into plenty of boiling salted water and cook 20 minutes. Drain and cool in ice water. Crack the shells; remove the meat. Pack into containers in recipe-size amounts, leaving ½-inch of head space. Seal, label and freeze.
Shrimp: Remove the heads, peel and devein, then rinse well in cold water. Bread, if you wish. Tray freeze just until solid, then pack in containers or bags, seal, label and freeze. Breaded shrimp do not need to be thawed before frying. Shrimp to be boiled do not need to be thawed; just add them to boiling water.
Oysters, Clams, Scallops: Shuck and wash them in salt water (1 gallon cold water and 1½ cups salt). Drain well. Pack in containers leaving ½-inch of head space. Seal, label and freeze.

Storing and Thawing

Store shellfish 2 to 3 months. Thaw shellfish in the refrigerator and cook as soon as possible once they are thawed.

Freezer Ideas

The star attraction of many a meal can wait in the freezer until you need it. When you are putting together a casserole, soup, stew or other specialty, double or triple the recipe and freeze half or two-thirds of it. It takes little extra effort on your part, but you get a dividend of one or two cooking-free afternoons or evenings.

Equipment for Main Dishes

Besides the cooking utensils needed for whatever main dish you intend to freeze, you will need plenty of heavy-duty aluminum foil, freezer tape, labels and a grease pencil or felt tip pen.

Ingredients

Almost any casserole, soup, stew or combination main dish freezes well: macaroni and cheese, baked beans, meat loaf, chop suey, chili, goulash, chicken a la king, meat or poultry pies or turnovers, Swiss steak, meat balls, hash, stuffed peppers, fried chicken, lasagne. Give your favorite recipes a try. If you run out of ideas, just take a stroll through the freezer section of the supermarket for inspiration.

Packaging

Undercook foods to be frozen and reheated; the reheating will complete the cooking. Macaroni products and rice, especially, must be undercooked or they will get mushy when reheated.

Cool prepared foods before packaging and freezing by putting a pan of food in ice water and stirring occasionally until cool.

Package main dishes in foil, so you can have your casseroles, baking dishes or pans free for everyday use. Start with an 8x8x2-inch or 9x9x2-inch square pan or your favorite

Casserole Freezing. Line dish with plastic wrap or foil. Spoon in recipe.

Wrap snugly and freeze solid. Then lift frozen food out of casserole and return food to freezer.

casserole. Tear off a sheet of heavy-or extra heavy-duty aluminum foil twice the size of the pan or dish. Put the pan or dish upside down on the counter, then press foil over bottom. Turn pan over, fit formed foil in and fill with food. Fold the foil loosely over food; freeze solid. Remove frozen food from pan, seal the foil with the Drugstore Wrap, tuck ends under and seal with tape. Label and return to the freezer.

Storing, Thawing and Cooking

Store frozen main dishes up to 3 months. Thaw frozen main dishes in the refrigerator for quicker reheating, if desired. Thawing is not necessary for most recipes, though.

Reheat or cook frozen prepared foods just as you would heat recipes normally, only cook them a little longer. Reheat main dishes in the oven, in a saucepan or skillet over medium heat (stirring to break up food as it thaws), in a double boiler or saucepan over boiling water, or in a microwave oven. Gravies and sauces may separate when reheating, but will smooth out if you stir them. Put cheese or crumb toppings on casseroles just before reheating.

Tray Dinners

Your freezer lets you be your own food processor, assembling dinners from leftovers ("plan-aheads" is really the better word) or cooking in quantity especially for freezing.

Equipment for Tray Dinners

Save aluminum trays from frozen dinners, buy divided aluminum trays in the housewares section, or form extra heavy-duty aluminum foil over the bottom of a plate, pie pan or small casserole. You will also need heavy-duty foil, plastic wrap or freezer paper, freezer tape, labels and a grease pencil or felt tip marker.

Tray Dinners. Save trays from commercially frozen dinners, or buy divided foil trays.

Packaging

To freeze extra servings from dinner, portion food onto foil trays just as you would serve the plates. Spoon meat juices, sauce or gravy over meat, sprinkle vegetables with a tablespoon of water and dot vegetables and potatoes with butter. Put the trays in the refrigerator to cool while you eat. After dinner, wrap the trays in foil, plastic wrap or freezer paper, label with contents, seal and freeze.

It is wise to include heating directions right on each package label: "375°F for 45 minutes," for instance. Fried foods should be uncovered for the last 10 to 15 minutes of heating so they can crisp.

If you cook foods just to tray freeze, undercook them slightly, especially the vegetables. Do not try to duplicate all the side dishes that come in frozen dinners — puddings, for example, will separate if made at home. Gravies and sauces may separate when reheated, but you can usually get them back in shape with a quick stir.

Dinner combinations can be anything your family enjoys. For starters, look at the list below, or take a look at the selection offered by commercial dinners. You will soon come up with popular tray dinners customized for your family.

You can cater to fussy eaters or special dieters by preparing several dinners for them at once, then freezing the dinners until needed.

Tray Dinner Possibilities

Swiss-steak, spinach and French fries
Fried chicken, carrots, lima beans, cranberry sauce
Meat loaf, whipped potatoes, beets, apple crisp
Sliced roast beef with gravy, scalloped potatoes, asparagus
Chicken a la king, biscuit, green beans, baked apple
Baked beans, hot dogs, zucchini, applesauce
Ham steak, candied sweet potatoes, French-style green beans
Ravioli, Italian green beans, baked pears
Turkey, dressing, sweet potatoes, mixed vegetables
Beef stew, garlic bread, apple Betty
Pot roast, peas and carrots, roast potatoes

Reheating

While it is very convenient to reheat everything all at once on a tray, some diners prefer the flavor of food heated separately. To do this without sacrificing too much convenience, group dinner-size packages of meat, vegetables, fruit and bread together in a large plastic bag or even a brown grocery bag. Label the meal bag with eating directions for each menu item. If the head cook is going to be away, name and locate refrigerator or pantry items to round out the meal. Only nonreaders can complain they were left at home with nothing to eat!

Do not forget to mark the meals you put in the freezer on your inventory, so you will know what is there. Dinners in the freezer are a cook's security blanket.

Sandwiches

Any lunch packer knows a trip to the freezer is much easier than making sandwiches morning after morning. It is really no more trouble to make a dozen or half-dozen sandwiches than two or three. Set up an assembly line once or twice a month and get the whole thing over with.

Equipment and Ingredients

You will need plastic wrap, heavy-duty foil, freezer paper or plastic sandwich bags, freezer tape, labels, a grease pencil or felt tip marker and a large plastic bag.

Mayonnaise and lettuce, two sandwich staples, do not freeze well, but there is an easy way around that problem. Prepare the sandwiches without lettuce or mayo, then when the sandwiches are ready to be packed in a lunch box, wrap the lettuce or mayo separately to tuck into the sandwich at lunch time. Anything else goes in frozen sandwiches. You can use bottled French or Russian dressing, cream cheese or cheese spreads instead of mayonnaise to hold the fillings together.

Always spread bread slices (you can use frozen or fresh bread) with softened butter or margarine right to the very edge. This butter-coating helps keep the bread from getting soggy. Top half of the bread slices with filling, then put on the second buttered bread slices.

Packaging

Wrap each sandwich individually in plastic wrap, foil, freezer paper or plastic sandwich bags. Heat-seal plastic bags with a sealing appliance, if you have one, or use zipper-type bags or freezer tape to seal bags. Label each sandwich. Pack all of one kind together in a box or large plastic bag, then freeze.

Storing and Thawing

Store sandwiches for a month, at 0°F, of course. Thaw sandwiches unopened. Put them right from freezer into lunch boxes or bags in the morning and they will be just right to eat at noon.

Sandwich Suggestions

Bread does not always have to be white. Try whole wheat, dark or light rye, egg, brown, date-nut, cinnamon, raisin, pumpernickel, onion, buns, rolls, even biscuits.

Fillings can be tuna, cheese, roast beef, corned beef, ham, bacon, sausage, crab, shrimp, chicken, liverwurst, chopped hot dog and cheddar cheese, ham salad, shredded cheddar and green pepper, nut or olive and cream cheese, peanut butter plus bacon, pickle or jelly.

Appetizers

While you are thinking about bread and sandwich fillings, transform some of these ingredients into bite-sized canapes or appetizers. Cut day-old bread into fancy shapes, top with the filling of your choice. Freeze on a tray, then package carefully in rigid containers or boxes. Seal, label and freeze. They will last a month or two. Unwrap, arrange them on a tray and thaw them in the refrigerator for an hour or so before you need them.

Dairy Products

Milk: Tuck away a quart or half-gallon of milk in a corner of the freezer, just for emergencies. Cream, both light cream for coffee and whipping cream, and milk can wait up to a month in the freezer. Thaw them in the refrigerator. Buttermilk, yogurt and sour cream will separate when thawed, but a spin in the blender will restore their smoothness.

Cottage Cheese: Cream-style cottage cheese will separate when thawed, but if you are going to sieve or blend it for salads or cheesecake, separation does not matter. Dry cottage cheese, farmer's or Ricotta cheese can be frozen for a month. Thaw them in the refrigerator.

Whipping Cream: Freeze it in liquid form, let it thaw in the refrigerator before whipping it. For better results, whip the cream and sweeten it as usual. Drop blobs of whipped, sweetened cream onto waxed paper on a cookie sheet and tray-freeze until solid. Then carefully transfer them to a freezer container or box, overwrap and store in a 0°F freezer for a month. Put a mound of frozen whipped cream on each dessert serving before you sit down to dinner and it will be thawed in time for dessert.

Cheese: Cheese freezes very well, as long as you realize the limitations. Frozen cheese will be crumbly, perhaps even a lit-

Freezing Whipped Cream. Sweeten as usual, then drop blobs onto foil or waxed paper-covered cookie sheet.

tle dry, and it will not slice perfectly, but the flavor will be just as good as fresh. Freeze cheese in small pieces — no more than an inch thick and no more than a half-pound per chunk. Seal it in foil, freezer wrap, plastic film or bag. Store cheese in a 0°F freezer for up to 6 months. Thaw it in the refrigerator and use as soon as possible after thawing.

Shredded cheese can be frozen, too, saving you a few minutes at cooking time. Shred cheese, spread the shreds in a thin layer on a cookie sheet and tray-freeze. Seal the frozen shreds in a bag or container and store them in the freezer, then just pour or spoon out the amount you need.

Butter or Margarine: Leave butter and margarine in their cartons. The cartons need an overwrapping of foil, freezer paper, plastic wrap or bag. Both butter and margarine are good up to 9 months in a 0°F freezer, so stock up when prices are low and you will have some on hand for baking or just for security. Thaw packages in the refrigerator or microwave oven (follow the manufacturer's directions). Butter pats, balls, curls or molds can wait in the freezer for special events. Tray-freeze, then package and seal them carefully to save their pretty shapes.

Ice Cream (and related products): As long as you close the package well, ice cream can stay in its original container. But if it is in a cardboard carton and has been opened several times, it will keep better if overwrapped with foil, plastic wrap, freezer paper or a plastic bag — or slip it into a container with a cover. After you have opened ice cream and scooped some out, prevent ice crystals on the surface by covering scooped portion with a piece of plastic wrap, foil or freezer paper before closing the carton. One month is maximum storage time for ice cream, ice milk, sherbet, etc. You can make and freezer-store your own ice cream specialties: sandwiches, cones, cakes, molds, pies. Take a look at some of the fancy creations next time you visit an ice cream shop, then duplicate those items at home.

Eggs

You can freeze eggs whole (broken and in containers, not in their shells), or as separate whites and yolks. They can be packaged in recipe-size containers or as one-egg cubes. Stock up when prices are low, then freeze so you will

Freezing Eggs. You can freeze whole eggs, egg whites or yolks to be used for omelets and main dishes.

always be able to whip up a cake, custard, souffle, omelet, cookies, or whatever.

Fresh, perfect eggs are the only ones that should go in the freezer. Break the eggs one at a time into custard cups. Check for freshness and quality and remove any bits of shell.

Whole Eggs: Break as many eggs, one at a time, as you need. Stir them gently with a fork to blend but do not beat in any air. Press the eggs through sieve or food mill. Measure and stir in ½ teaspoon salt for each cup of eggs to be used for scrambled eggs, omelets, souffles; stir in 1½ teaspoons corn syrup to each cup of eggs to be used in cakes, cookies or other sweet cooking. Measure 2½ to 3 tablespoons for each egg and package recipe-size amounts in containers or plastic bags or pouches. To freeze one-egg cubes, check to see how much a single section of your ice cube tray will hold. If that measurement is close to 3 tablespoons, figure one egg per cube. Break and prepare the eggs as above and pour the eggs into the tray. Freeze solid, then transfer the frozen cubes to a bag or container, seal and label, indicating sweetened or salted and freezer-store. If your ice cube sections are smaller, work it out so that two cubes will equal one egg. Freeze, then package and freezer store.

Egg Yolks: Break the eggs, one at a time, as above, and separate the whites from the yolks. Stir the yolks gently with a fork but do not beat in any air. Press them through a sieve or food mill. Measure and stir in ½ teaspoon salt or 1 tablespoon sugar or corn syrup (depending on their final use) for each cup of yolks. Measure 1 tablespoon for each egg yolk and freeze in containers, bags or pouches, or measure 1 tablespoon into each section of an ice cube tray. Freeze, then transfer the frozen yolks to plastic bags or containers and freezer-store. Be sure to label the packages with the number of egg yolks and whether they are packed with salt or sugar.

Egg Whites: Break the eggs, one at a time, as above, and separate. Press the whites through a food mill or sieve, then measure 1½ to 2 tablespoons for each white. Package in recipe-size amounts in containers, bags, pouches. Seal, label with number of egg whites, and freeze.

Storing and Thawing

Store frozen eggs 9 to 12 months at 0°F. Thaw frozen eggs, unopened, in the refrigerator, never at room temperature. Never refreeze thawed eggs.

Salads

Salads made with a cream cheese or a whipped cream base freeze beautifully. Some gelatin-based salads freeze well, too. Just remember to use a little less water than usual, so the gelatin will be stiffer.

Fruit salads made with cream cheese or whipped cream can be frozen in large or individual molds, in fruit or juice concentrate cans, milk cartons or even paper-lined muffin cups. Use small pieces of fruit in these salads for easier eating. The fruit freezes hard and eating the large pieces can be like eating ice cubes. A month is maximum storage time for most salads. Frozen fruit salads should be thawed only slightly before serving; gelatin salads can be thawed in the refrigerator. Gelatin salads can be put in the freezer compartment to make them set faster — but do not forget them.

Tossed salads and fresh vegetable salads do not freeze well, but they take only minutes to put together, anyway.

Baked Products

With the exception of cream, custard, or meringue pies, or cakes with cream or custard fillings, most other baked goods take beautifully to the freezer. Bake when you feel the urge, then let the freezer hold the fruits of your labor. Or, buy when the price is right or when you can get to a day-old baked goods store or outlet store. Then you always will have something on hand for after school, impromptu cake and coffee sessions, or bake sales. For packaging you will need heavy-duty foil, plastic wrap, freezer paper, or plastic bags, ties for the bags, freezer tape, labels, grease pencil or felt tip marker.

Bread: Keep commercial frozen bread dough on hand, make up your own from the specially formulated recipes that follow, bake your own bread to freeze, or pack away bakery or store bread. Wrap baked bread well in foil, plastic wrap, freezer paper or bag and seal. Store bread will be alright for a week in its original wrappings. For longer storage than that, overwrap the bread or bag it. Baked bread can stay in the freezer for 6 months at 0°F, unbaked bread should only be stored for 1 month. Follow the recipe's or label's directions to thaw the bread dough. Baked bread can be thawed unopened at room temperature or in non-plastic wrappings in 325°F oven for about 20 minutes. It is wise to slice home-baked bread before you freeze it, then you can take out only what you need. Single slices thaw quickly at room temperature or in the toaster.

To make your own "Brown and Serve Rolls," bake yeast rolls at 325°F for 15 minutes. Cool, then package, seal, label and freeze. To serve, heat in a 450°F oven about 10 minutes or until nicely browned.

Quick breads, sweet rolls, pancakes, waffles, muffins, popovers, doughnuts, all can be wrapped or bagged and sealed, then freezer stored for up to 4 months. Wrap the items individually if you have a family that likes to snack. Thaw the frozen baked goods in wrappings at room temperature, or in a 325°F oven, toaster-oven or toaster, depending on size of bread.

When you make garlic bread, fix a few extra loaves; wrap, seal, label and freeze the extras.

Bread crumbs for toppings, coatings, and stuffing can be freezer-stored, as can cubes of bread for croutons, puddings or stuffing. Cook additional stuffing at holiday time and freeze the surplus.

Cakes: All cakes freeze well — layer, loaf, cupcakes, angel, chiffon, sponge and fruitcakes. Cool thoroughly after baking and freeze them unfrosted or frosted. Wrap or bag-seal individual unfrosted layers or cakes. Thaw them at room temperature in the wrappings for about an hour, then frost. Egg white frostings do not freeze well. Choose butter, fudge or caramel frostings to top cakes and be frozen. Put the cakes on circles of heavy cardboard, frost, and decorate if you wish. If you frost cakes before freezing them, freeze them unwrapped until the frosting is hard. Carefully cover the cakes with plastic wrap, foil or a large bag and seal. Put frosted cakes in a bakery box or plastic cake carrier to protect them from being scrunched in the freezer. Cupcakes are great to have on hand in the freezer.

Unfrosted cakes can stay in the freezer 6 months at 0°F; frosted cakes can be stored for 2 months. Angel and sponge cakes are best used within 1 month. Thaw frosted cakes unwrapped, but under a cake cover, to prevent condensation on frosting.

Some cooks like to freeze cake, pancake, muffin or waffle batter. The results are risky and less than good. It takes less time to mix up a batch of batter than it does to thaw frozen batter, anyway.

Cookies: Cookie lovers always save some space in the freezer for cookie dough so they can bake batches when they feel like it. Baked cookies can also be frozen. Busy Christmas cookie makers start their mixing and baking in the fall, then freeze baked cookies or dough until the holidays. You have to be discreet about how and where you put cookies to be saved, though. One smart cook, with several teenagers, labels Christmas cookies "Liver."

Package baked cookies in freezer containers, shortening or coffee cans, plastic food storage containers or plastic bags. They will keep in the freezer up to a year, if well hidden. Remove them from the containers to thaw — they take only a few minutes.

Rolled cookie dough can be frozen in several large flat patties, placed between sheets of freezer wrap, plastic wrap or foil and sealed in a plastic bag. Thaw them in the refrigerator for an hour or so before rolling and cutting. Another way to freeze cookie dough is to roll out the dough, cut the cookies and tray-freeze the cutouts on waxed paper or cookie sheets. When frozen solid, carefully stack the cutouts in containers or boxes with waxed paper or freezer paper in between. Bake without thawing.

Drop cookie dough can be shaped into rolls and wrapped well in plastic wrap or foil to be frozen. Slice and bake the cookies without thawing. To tray-freeze the dough, drop tablespoons of it onto cookie sheets, freeze them until solid, then layer the drops in containers or boxes with freezer paper or waxed paper between the layers. Bake without thawing.

Cookie doughs can be stored in the freezer up to a year.

Pies: You can freeze pies baked or unbaked. Fillings and pastry or crumb crusts can also be frozen separately. Fruit pies are usually best frozen before baking. Cheese or pecan pies or others with sticky fillings are better baked, then frozen. Baked pies can be cut and the pieces frozen individually to go into lunchboxes.

Unbaked pies (single or double crust) are prepared for the freezer just as if you were going to bake them right away, but do not cut vents in the top crust. Invert a paper plate over top

Protect the edge of unbaked pies by cutting strips of cardboard to wrap around pie. Tape shut.

of each pie to protect the crust, then wrap it in heavy-duty foil, freezer paper or plastic wrap, seal, label and freeze. Freeze pies before wrapping, if you wish. Protect pies from being damaged in the freezer by reserving a protected corner for them, or by wrapping each with a cardboard collar or putting the pie in a bakery box. Unbaked pies can be stored in a 0°F freezer for 3 to 4 months. To bake, unwrap, cut vents in top crust and bake as usual, adding 15 to 20 minutes to baking time.

Baked pies should be completely cooled before freezing. Protect the tops with paper plates, then wrap, seal, label and freeze for a month or two. Thaw at room temperature for at least an hour or heat in 375°F oven for 30 to 45 minutes.

Chiffon pies should be chilled in the refrigerator until set, then frozen solid. Wrap them carefully with heavy-duty foil or plastic wrap. Protect them with a cardboard collar, plastic pie

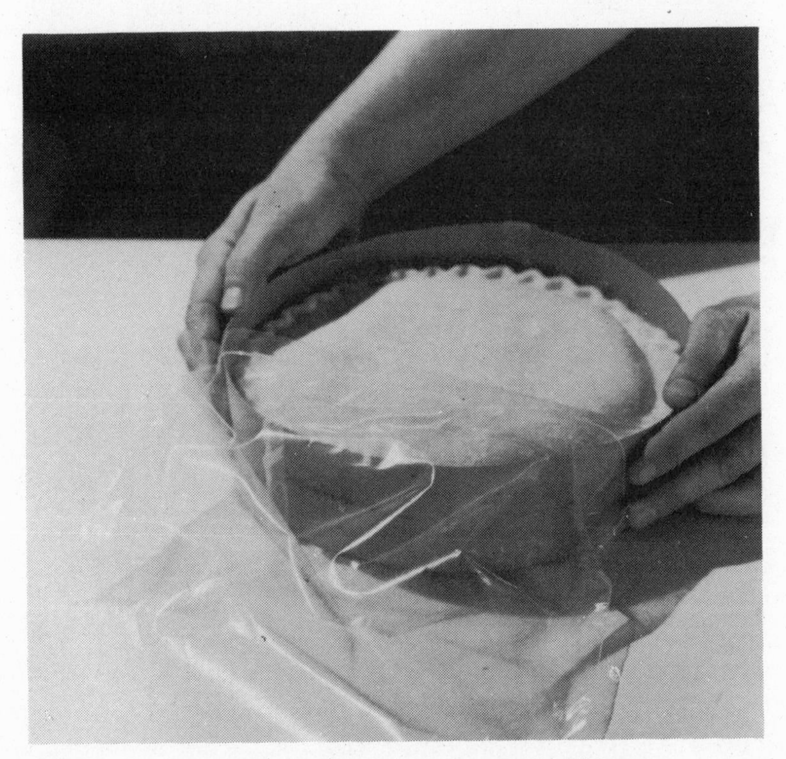

Slip pie in cardboard collar into plastic bag, seal and freeze.

holder or bakery box. Chiffon pies can be stored in a 0°F freezer for a month or two. Thaw them in the refrigerator for a few hours, or serve frozen.

Pastry can be frozen in one-crust portion patties. It can also be frozen in rolled out circles, or in pans or pie plates. Rolled out circles can be stacked on a larger circle of heavy cardboard, with two thicknesses of waxed paper or freezer paper between each circle. Wrap or bag, seal, label and freeze. Remove as many circles as you need and let them thaw at room temperature until pliable, then fit them into the pie plates or pans. Bag and seal one-crust portion patties, then freeze. Thaw them at room temperature, then roll out. Pastry shells are best frozen in foil or paper pie plates so you can have your metal or glass pans for day to day use. Wrap each pastry shell in foil, plastic wrap or put each shell in a plastic bag, seal, label and freeze. If you are careful, you can stack the shells with a layer of crumpled waxed paper or plastic wrap between each one. Protect the shells in a plastic container, bakery box or with cardboard collars. Pastry can be stored in a 0°F freezer for 4 to 6 months.

Freezer Tips

When you learn to live with a freezer, you are bound to come up with lots of freezer ideas of your own to add to this list.

Beverages: Almost any leftover non-alcoholic beverage can be frozen in ice cube trays. Add the cubes to that same beverage (juice, coffee, tea, punch) and you can enjoy it cold but not diluted. Freeze syrup from canned fruit, too.

Candied and dried fruits, crystallized ginger, ginger root, coconut, grated or shredded citrus peel wait in the freezer until you need them.

Candies, homemade or otherwise, can stay in the freezer instead of turning stale on the shelf. Wrap individual pieces or bag them, seal and freeze.

Cereals and snacks: Cereal, potato chips, popcorn, even pretzels stay fresh and crisp in the freezer.

Coffee: Keep coffee tightly covered and in the freezer for maximum flavor.

Cream puffs: Bake cream puffs, fill them with whipped or ice cream, wrap, seal, label and freeze for an elegant occasion.

Dips: Making dips for a party? Double the batch and freeze half. Thaw them in the refrigerator. Whip or blend in blender to smooth out, if necessary.

Flours and nuts: Whole grain flours and meals, nuts in and out of shell, seeds, stay fresh and safe in the freezer.

Frozen treats for kids are much cheaper if you create them yourself. Use paper cups and wooden sticks, then fill the cups with apple cider, pineapple or grape juice, fruit cocktail, yogurt, canned fruit drinks, reconstituted packaged fruit drinks. Add a stick to each cup before freezing for a handle. Alternate layers of ice cream, nuts, sundae sauces in paper cups for portable parfaits.

Half-empty cans of olives, water chestnuts, tomato paste, pineapple, kraut, bamboo shoots can wait in the freezer until you need them, rather than rotting in the refrigerator.

Ice is much cheaper when you make it. Empty ice cube trays into a plastic bag or containers whenever you think of it and build up a reserve. Freeze ice rings for punch bowls, putting flowers, fruit, leaves in water before freezing. Make fancy cubes the same way. Freeze water in milk cartons to use for blanching and homemade ice cream. Freeze some water in a sealed, small, boilable pouch to keep on hand for a mini ice pack.

Leftover bacon: Just crumble, package and freeze to use later on in sandwiches, recipes, for toppings or salads or baked potatoes.

Marshmallows and brown sugar stay soft in the freezer. Frozen marshmallows are easy to cut, too.

Pizza: Put together your favorite pizza, wrap, seal and freeze it to bake another day. Reheat as usual, adding about 10 minutes baking time.

Sauces: Even the fanciest sauces for meats, vegetables, desserts can go in the freezer in containers or boilable pouches. Reheat them slowly, stirring occasionally. Make cream or white sauce in a bigger than usual batch, then pour it into ice cube trays to freeze. Transfer the cubes to bags or containers and store them in the freezer. Toss a cube into cooked vegetables for quick creamed vegetables.

Soup: Soup aficionados say flavor is better after freezing. Cook up a big pot, then freeze it in family-size portions to eat later on. Stash soup bones, well wrapped, in the freezer until you are ready to make soup.

Recipes

DINNER ROLLS FOR THE FREEZER

You can impress short-notice guests with home-made rolls! These rolls thaw and rise in an hour and a half and bake in 15 minutes. Makes about 4 dozen rolls.

Ingredients

- 5½ *to 6½ cups unsifted all-purpose flour*
- ½ *cup sugar*
- 1½ *teaspoons salt*
- 2 *packages dry yeast*
- 1¼ *cups water*
- ½ *cup milk*
- ⅓ *cup margarine or oil*
- 2 *eggs (at room temperature)*

Equipment

- *Electric mixer with large bowl*
- *Small saucepan*
- *Measuring cups*
- *Measuring spoons*
- *Rubber spatula*
- *Kneading board*
- *Plastic wrap*
- *Tea towel*
- *Baking sheets*
- *Foil*
- *Plastic bags*
- *Cooling racks*

1. In the bowl of an electric mixer combine 2 cups of the flour, the sugar, salt and yeast; mix well.
2. In small saucepan, combine the water, milk and margarine or oil and heat over low heat until very warm (120-130°F). The margarine does not need to melt.
3. With the mixer running, gradually add the hot mixture to the dry ingredients and beat 2 minutes at medium speed, scraping the bowl occasionally.

4. Add the eggs and ½ cup of the flour and beat at high speed 2 minutes, scraping bowl occasionally.
5. Stir in enough of the remaining 3 to 4 cups flour to make a soft dough.
6. Turn out the dough onto a lightly floured board or countertop and knead about 8 to 10 minutes or until smooth and elastic. Cover it lightly with plastic wrap and a tea towel and let it rest 20 minutes.
7. Punch down the dough. Cut the dough in half, then cut each half into 24 equal pieces. Shape as directed below or create your own favorite shape.
   Spirals: Roll each section to a rope about 9 inches long. Hold one end of rope in place on greased baking sheet and wind closely to form a coil. Tuck the end underneath to hold it in place.
   Quick Cloverleafs: Roll each section into a small ball and place in a greased muffin tin. With scissors, cut each ball in half, then in quarters, cutting almost to bottom of dough.
   Knots: Roll each section into a rope about 9 inches long. Loosely tie each section once and put the knots on greased baking sheets.
8. Cover the rolls on the baking sheets or in the muffin tins with plastic wrap and then with foil; seal them.
9. Freeze solid. When solid, transfer the rolls to plastic bags and freeze up to 4 weeks.
10. To Thaw: Remove the rolls from the freezer and from the plastic bags. Place them on greased baking sheets. Cover lightly with plastic wrap and let them rise in a warm draft-free place until doubled, about 1½ hours.
11. Bake in a preheated 350°F oven about 15 minutes or until golden brown. Remove the rolls from the baking sheets and serve at once.

WHITE BREAD FOR THE FREEZER

This recipe is especially formulated to be frozen up to one month. Do not try to freeze ordinary bread recipes; they probably will not work. Makes 4 (8½x4½-inch) loaves.

(Continued On Next Page)

Ingredients	Equipment
12½ to 13½ cups unsifted all-purpose flour	Electric mixer
½ cup sugar	Large mixer bowl
2 tablespoons salt	Rubber spatula
⅔ cup instant nonfat dry milk	Measuring cups
4 packages dry yeast	Board
¼ cup softened margarine	Baking sheets
4 cups very warm (120-130°F) tap water	Plastic wrap and bags
	2 (8½x4½-inch) loaf pans
	Cooling racks

1. In the bowl of an electric mixer combine 4 cups of the flour, the sugar, salt, dry milk and yeast; mix well and add margarine. With the mixer running, gradually add water, then beat 2 minutes at medium speed, scraping the bowl occasionally.
2. Add 1½ cups flour and beat at high speed 2 minutes, scraping the bowl occasionally. Stir in enough of the remaining 7 to 8 cups of flour to make a stiff dough.
3. Turn the dough out onto a lightly floured board or countertop and knead about 15 minutes or until the dough is smooth and elastic. Cover lightly with plastic wrap and let it rest 15 minutes.
4. Cut the dough into fourths and form each quarter into a smooth round ball. Flatten each ball until it is about 6 inches in diameter. Put the 4 balls on 1 large or 2 small greased baking sheets.
5. Cover the flattened balls with plastic wrap and freeze solid.
6. When frozen solid, quickly transfer them to plastic bags. Seal or tightly close the bags and freeze the dough up to 4 weeks at 0°F.
7. To Thaw: Remove the balls one at a time, or all at once, from the freezer and plastic bag. Place on an ungreased baking sheet. Cover them lightly with plastic wrap and let them stand at room temperature about 4 hours or until fully thawed.
8. Round Loaves: Let the dough rise on baking sheets in a warm, draft-free place until doubled in size, about 1 hour. Bake as directed below.
   Regular Loaves: Roll each ball of dough into a rectangle 12x8 inches. Roll up the rectangle jelly roll fashion, start-

ing with the 8-inch side. Pinch the seam to seal, then seal the ends by pressing them with the side of your hand or pinching the ends with your fingers. Put each loaf into a greased 8½x4½x2½-inch loaf pan. Let them rise in warm draft-free place until doubled in size, about 1½ hours.

9. Bake in a preheated 350°F oven about 35 minutes or until the loaf sounds hollow when tapped and is nicely browned. Remove the loaves from the pans and cool on racks.

RYE BREAD FOR THE FREEZER

Replace the caraway seed with shredded orange peel, if you like. You could also substitute honey for the molasses, or, if you prefer, use whole wheat flour in place of rye flour. Makes 2 round loaves.

Ingredients	Equipment
3½ to 4½ cups unsifted all-purpose flour	*Electric mixer with a large bowl*
2 cups unsifted rye flour	*Measuring cups*
1 tablespoon salt	*Measuring spoons*
1 tablespoon caraway seed	*Rubber spatula*
2 packages dry yeast	*Kneading board*
¼ cup softened margarine	*Baking sheets*
2 cups very warm (120-130°F) tap water	*Plastic wrap*
	Plastic bags
⅓ cup molasses	*Cooling racks*

1. In the bowl of an electric mixer combine 1 cup of the all-purpose flour, 1 cup of the rye flour, the salt, caraway seed and yeast; mix well and add the margarine.
2. With the mixer running, gradually add the water and molasses, then beat 2 minutes at medium speed, scraping the bowl occasionally.
3. Add ¾ cup of the all-purpose flour and beat at high speed 2 minutes, scraping bowl occasionally. Stir in the remaining 1 cup rye flour and enough of the remaining 1¾ to 2¾ cups all-purpose flour to make a stiff dough.
4. Turn out the dough onto a lightly floured board or coun-

(Continued On Next Page)

tertop and knead about 8 to 10 minutes, or until the dough is smooth and elastic.

5. Cut the dough in half. Form each half into a smooth round ball, then flatten each ball until it is about 7 inches in diameter. Put the flattened balls on greased baking sheets. Cover with plastic wrap and freeze solid.
6. When solid, quickly transfer them to plastic bags. Seal the bags and freeze the dough up to 4 weeks.
7. To Thaw: Remove the dough from the freezer and from the plastic bag. Place the rounds on an ungreased baking sheet. Cover them lightly with plastic wrap and let them stand at room temperature about 2¼ hours or until fully thawed. Then let them rise in a warm draft-free place until doubled in size, about 2¼ hours.
8. Bake in a preheated 350°F oven about 35 minutes or until loaves sound hollow when tapped and are nicely browned. Remove them from the baking sheet and cool on racks.

PEACHES AND CREAM CHEESE SALAD

You can freeze some gelatin-based salads and here is a quick-to-fix example. Makes 4 (8x8-inch pans) or 2 (9x13-inch) pans or a total of 24 servings.

Ingredients

2 packages (6 ounces each) lime-flavored gelatin
1 quart boiling water
2 cans (1 pound 13 ounces each) sliced cling peaches
2 cans (1 pound 4 ounces each) crushed pineapple
4 pounds cottage cheese

Equipment

Small saucepan or teakettle
Large mixing bowl
Measuring cup
Sieve or strainer
Mixing spoon

1. Line the pans with freezer paper, foil or heavy-duty plastic wrap, leaving enough wrap to fold over the sides and ends.
2. In a large mixing bowl, stir the gelatin and boiling water together until gelatin dissolves.

3. Drain the peaches and pineapple, reserving the syrup. Measure 2 cups syrup and add it to the dissolved gelatin.
4. Stir in the peaches, pineapple and cottage cheese.
5. Divide the gelatin mixture between the pans. Refrigerate until firm.
6. Fold the freezer wrapping over the gelatin salads and seal with the Drugstore Fold or tape.
7. Label and freeze. When frozen solid, lift the wrapped salad out of the pan and return it to the freezer so you will have the pan to use.
8. To Serve: Unwrap the salad and put it back in pan. Thaw it in the refrigerator about 24 hours. Cut the salad into squares and serve on crisp salad greens.

ALMOST ANYTHING CASSEROLE

Use chicken, ham, cooked beef, textured vegetable protein, cooked pork, tuna or just double the amount of cheese for this handy casserole. Makes 24 (2/3 cup) servings.

Ingredients

3 cups (12 ounces) uncooked elbow macaroni
1½ sticks (¾ cup) butter or margarine
1 cup flour
4 teaspoons salt
¼ teaspoon pepper
1 teaspoon marjoram, basil, poultry seasoning or thyme
2 quarts milk
1¼ quarts diced cooked turkey or other meat (see above)
10 ounces (2½ cups) shredded cheddar, Swiss, Muenster or Colby cheese
2 jars (4 ounces each) pimiento, drained and chopped

Equipment

Large saucepan
Measuring cups
Measuring spoons
Dutch oven or large kettle
Colander or strainer
Knife
Containers
Pans or casseroles
Heavy-duty foil, plastic wrap or freezer paper
Labels and marker

(Continued On Next Page)

1. Line the pans or containers with foil, freezer paper or plastic wrap, leaving enough wrap to fold over the sides and ends.
2. In a large kettle, heat 1 gallon water and 1 tablespoon salt to boiling. Add macaroni and cook about 6 minutes or until just tender. Drain.
3. Melt the butter in a large saucepan. Blend in the flour and cook and stir over medium heat until foamy. Stir in the seasonings.
4. Gradually add the milk. Cook and stir over medium high heat until the mixture comes to a boil and is smooth and thick.
5. Stir in all remaining ingredients. (If desired, save about half the cheese to sprinkle over the top.)
6. Divide the casserole among the lined containers in amounts that are best for your family. Fold over the wrap and seal with the Drugstore Fold or with tape.
7. Label and freeze until solid. If you want to have containers to use again right away, lift out the frozen food and return the food to the freezer.
8. To Serve: Unwrap and put the frozen casserole in a baking pan or other ovenproof container. Heat in a 350°F oven 1¼ hours or until hot through. Or, thaw and heat in a microwave oven, in a heavy saucepan over low heat, or in a double boiler.

CHICKEN A LA KING

When chickens are cheap, buy several to cook up for this freeze-ahead special. Freeze it in 8x8- or 9x13-inch pans, in foil pie pans, in casseroles, or in almost any convenient container. By lining a container with foil or plastic wrap you can have it back in use again as soon as the Chicken a la King is frozen solid. Makes 24 (1 cup) servings.

Ingredients

1 1/3 cups water
2 packages (10 ounces each) frozen green peas
1 cup chopped green pepper
½ cup chopped onion or 2 tablespoons instant minced or dried onion
2 quarts chicken broth
1 2/3 cups flour
1 quart milk
2 quarts diced cooked chicken
2 cans (8 ounces each) mushroom pieces, drained
1 jar (4 ounces) pimiento, drained and chopped
3 tablespoons salt
1 to 2 teaspoons poultry seasoning
Dash pepper

Equipment

Large saucepan
Measuring cups
Measuring spoons
Strainer
Mixing spoon
Knife
Freezer paper, heavy-duty foil or plastic wrap
Containers, pans or casseroles for freezing
Labels and marker

1. Line the pans or containers with foil, freezer paper or plastic wrap, leaving enough wrap to fold over the sides and ends.
2. Heat water to a boil in a large saucepan. Add the peas, green pepper and onion. Cover and simmer 5 minutes. Drain, reserving the cooking water. Return the cooking liquid to the saucepan along with the chicken broth and heat to boiling.

(Continued On Next Page)

3. Blend the flour and milk until smooth, then gradually stir them into the hot broth mixture. Cook over medium heat until smooth and thick, stirring constantly. Stir in all remaining ingredients, including the cooked pea mixture.
4. Divide among lined containers in amounts that are best for your family. Fold over the wrap and seal with Drugstore Fold or tape.
5. Label and freeze until solid. If you want to have containers to use right away, lift out the frozen food and return the food to the freezer.
6. To Serve: Unwrap the Chicken a la King and put it in an ovenproof container. Heat it in a 350°F oven about 1½ hours or until heated through, stirring occasionally. Or, thaw and heat it in a microwave oven, or heat it in a heavy saucepan over very low heat, or in a double boiler until heated through. Serve over biscuits, noodles, or rice.

LASAGNE, MID-WESTERN OR ITALIAN STYLE

When you feel like cooking, cook in quantity, then freeze lasagne to enjoy later on. Use cottage and cheddar cheese for Midwestern Lasagne, Ricotta and Mozzarella for Italian. Makes 4 (8x8-inch) pans or 2 (9x13-inch) pans or a total of 24 servings.

Ingredients

- 2½ *pounds lean ground beef*
- 1½ *cups chopped onion or ⅓ cup instant minced onion*
- 2 *cloves garlic, minced, or ¼ teaspoon garlic powder*
- 3 *tablespoons chopped parsley or dried parsley flakes*
- 1 *tablespoon basil or oregano*
- 1 *tablespoon plus 1 teaspoon salt*
- 3 *cans (8 ounces each) tomato sauce*
- 3 *cans (6 ounces each) tomato paste*
- 2¼ *cups hot water*
- 3 *eggs, beaten*
- 3 *pounds cottage or Ricotta cheese*
- ¾ *cup grated Parmesan cheese*
- 2 *packages (12 or 16 ounces each) wide lasagne noodles*
- 1½ *to 2 pounds shredded cheddar or Mozzarella cheese*

Equipment

- *4 (8x8x2-inch) baking dishes or pans or 2 (9x13-inch) baking dishes or pans*
- *Heavy-duty foil*
- *Large skillet or Dutch oven*
- *Measuring spoons*
- *Medium mixing bowl*
- *Large saucepan or kettle*
- *Colander or strainer*
- *Labels and marker*

1. Line the baking pans with foil, leaving enough hanging over the sides of the pan to fold over the finished lasagne.

(Continued On Next Page)

(If baking one pan to eat right away, do not line it with foil but grease it lightly.)

2. Brown beef in a large skillet or Dutch oven, stirring to break up the meat. Add the onion and garlic and cook until the onion is tender. Drain off excess fat.
3. Stir in the parsley, basil and salt along with the tomato sauce, tomato paste and water. Simmer 20 to 30 minutes, stirring occasionally.
4. In a medium mixing bowl, beat the eggs with cottage (or Ricotta) cheese and Parmesan cheese.
5. Cook the lasagne noodles in boiling salted water (2 gallons water plus 2 tablespoons salt and 2 tablespoons oil) until tender but still firm, about 14 minutes. Drain well.
6. For 4 (8x8-inch) pans: Layer the ingredients in each of the 4 pans as follows — ¾ cup meat sauce, 2½ to 3 lasagne noodles, ¾ cup cottage cheese-egg mixture, ⅓ to ½ cup Mozzarella or Cheddar cheese. Repeat the layers until all ingredients are used.

   For 2 (9x13-inch) pans: Layer the ingredients in each of the 2 pans as follows — 1½ cups meat sauce, about 5 or 6 lasagne noodles, 1½ cups cottage cheese-egg mixture, ⅔ to ¾ cup Mozzarella or cheddar cheese. Repeat layers.
7. To Eat Right Away: Bake in a preheated 400°F oven about 30 minutes. Let it stand 10 minutes before cutting.
8. To Freeze: Fold up the foil hanging over the pan, fold the edges together to seal; seal the ends of the foil by folding them together. Label and freeze solid. Lift the foil-wrapped lasagne out of the pan when frozen so you will have the pan to use.
9. To Bake: Unwrap the lasagne and put it back in its pan. Bake in a 400°F oven 1 hour or until the center is hot. Let it stand about 10 minutes before serving.

Glossary

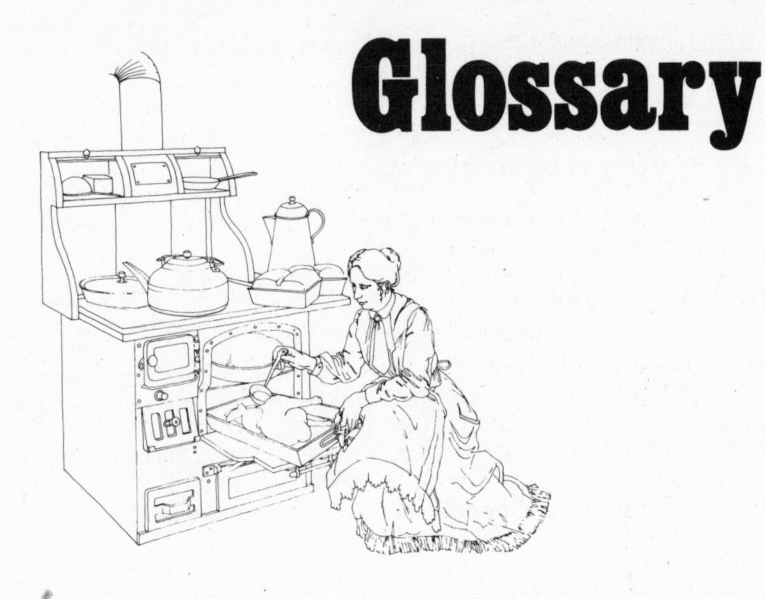

ASCORBIC ACID. Vitamin C, used to prevent fruits from darkening.

BACTERIA. Microorganisms that can grow in food and cause spoilage.

BLANCHING. Briefly pre-cooking vegetables to kill enzymes.

BOILING WATER BATH. Processing foods in enough boiling water to completely cover jars.

BOTULISM. Food poisoning caused by toxin produced in food by Clostridium botulinum. This toxin is one of the deadliest known to man.

BRINE. Salt-water solution used in pickling.

BUTCHER WRAP. A special wrap used for freezing foods, especially meat.

BUTTER. Very thick sweetened and spiced fruit puree.

CAN. To preserve food by sealing in jars and then processing to kill microorganisms that cause spoilage.

CHUTNEY. A highly spiced mixture of fruits and/or vegetables, sugar and vinegar; a condiment.

CITRIC ACID. A natural acid used to increase the acidity of low-acid tomatoes or to help prevent darkening of light-colored fruits. Available from drugstores.

COLD OR RAW PACK. Packing fruit or vegetables in hot canning jars without any pre-cooking.

CONSERVE. Jam made from one or more fruits, usually with raisins or nuts added.

CROCK. A large cylindrical pottery, earthenware or stoneware container to hold pickles or kraut during fermentation.

DRUGSTORE WRAP OR FOLD. A special over and over fold used to close packages for freezing.

FREEZE. To preserve foods by holding them at 0°F.

FREEZER BURN. Dry, tough, brownish areas on food that has been exposed to air during freezing; caused by ice crystals that evaporate from surface of food.

HEAD SPACE. The amount of space left between food or liquid in jar or freezer container and the top of a jar or container. Needed for food to expand during heat processing or freezing.

HOT PACK. Pre-cooking fruits or vegetables before packing them into hot canning jars.

JAM. Soft, sweetened fruit mixture with crushed or chopped fruit throughout.

JAR. Specially tempered glass container used to hold food for canning. May be wide-mouthed or have tapered sides.

JELLY. Sweetened fruit juice cooked with or without added pectin until it jells, or will hold a shape.

LID. Top for container; particularly flat metal top, rimmed with sealing compound, used to seal glass canning jars.

MASON JAR. A standard glass canning jar with a threaded top; named for its inventor, John L. Mason.

MARMALADE. Jelly with bits of fruit throughout; usually made from citrus fruits.

MOISTURE-VAPOR-PROOF. Impervious to water, air or odors. Necessary for proper protection of foods being wrapped or packaged for freezing.

OPEN KETTLE. Cooking food in an open kettle, then packing into sterile jars and sealing. The open kettle method is not recommended for any foods other than jellies.

PACKING. Putting fruits or vegetables into jars or freezer containers. Also, the material used to surround root vegetables for cold storage.

PARAFFIN. Household wax used to seal jelly in glasses.

PECTIN. A water soluble substance found in many fruits, used to form jellies. It comes in liquid or powdered form.

PICKLE. To preserve food with vinegar or brine. Also, a cucumber or other fruit or vegetable that has been preserved in vinegar or fermented in brine.

PRESERVE. To prevent food from spoiling. Also whole pieces of fruit simmered in a sugar syrup until plump and tender.

PRESSURE CANNER. A special cooking container that heats under pressure and at temperatures above boiling.

PROCESSING. Heating filled jars of food in a boiling water bath or pressure canner at 10 pounds of pressure for specified amounts of time.

REHYDRATION. Adding water to dried food to replace moisture lost during drying.

RELISH. Any preserved or pickled fruit or vegetable, usually chopped and highly spiced with vinegar added.

SALT. Sodium chloride. For pickling, pure granulated salt is preferred to regular table or iodized salt.

SAUERKRAUT. Shredded cabbage fermented in brine.

SCREW BAND. A metal threaded rim that screws on a standard glass canning jar and holds self-sealing lid in place.

SEAL. To close so that air cannot get in or out. In canning, sealing is done with lids; in freezing, sealing is done by wrapping or closing in containers with tight-fitting lids.

SHEET TEST. A test used to see if jelly is done. Sweetened juice is cooked until it flows from spoon in a sheet or flake.

SPOILAGE, SIGNS OF. Bulging or unsealed lid, foaming or mold on food, spurting liquid when a container is opened, off-odor, slimy or slippery food, off-flavors.

STERILIZING. To boil jelly glasses for 15 minutes.

SYRUP. Sugar-water solution used to sweeten fruits for canning and freezing.

THAW (OR DEFROST). To bring frozen food to room temperature.

VINEGAR. Acid liquid used in pickling and preserving. Cider vinegar is made from apple juice; white distilled vinegar is made from grain alcohol. Vinegar for pickling must be 4% to 6% acidity.

Recipe Index

A

D

E

F

J

L

M

N

O

P

R

T

U

V